Thermal and Time Stability of Amorphous Alloys

Thermal and Time Stability of Amorphous Alloys

A. M. Glezer
A. I. Potekaev
A. O. Cheretaeva

CISP

CRC Press
Taylor & Francis Group
Boca Raton London New York

CRC Press is an imprint of the
Taylor & Francis Group, an **informa** business

CRC Press
Taylor & Francis Group
6000 Broken Sound Parkway NW, Suite 300
Boca Raton, FL 33487-2742

First issued in paperback 2020

© 2017 by CISP
CRC Press is an imprint of Taylor & Francis Group, an Informa business

No claim to original U.S. Government works

ISBN-13: 978-0-367-57316-4 (pbk)
ISBN-13: 978-1-138-06827-8 (hbk)

Visit the Taylor & Francis Web site at
http://www.taylorandfrancis.com

and the CRC Press Web site at
http://www.crcpress.com

Contents

Foreword

The efforts of material scientists have resulted in the development of a large variety of structural states of materials characterised by unique physical–chemical and mechanical properties. However, all these materials, with a small number of exceptions, are far away from the thermodynamic equilibrium state. Therefore, it is not surprising that in the process of long-term or even short-term destabilising effects (temperature, deformation, radiation, etc) the structural–phase states of highly non-equilibrium materials transform in accordance with the general laws of thermodynamics and lose their unique properties. This circumstance greatly restricts or even completely prevents the practical application of highly non-equilibrium materials with unique properties in technology and medicine.

Undoubtedly, these materials with the non-equilibrium structure and unique physical–mechanical properties include metallic alloys produced by melt quenching at a rate exceeding 1 million degrees per second. Under the destabilising effects (especially thermal effects) the structure of these materials undergoes structural and phase transformations which greatly reduce the efficiency and often prevent application in practice. For example, the iron-based amorphous alloys characterised by unique magnetically soft properties combined with high strength, ductility and wear resistance are subject to stringent requirements to prevent changes of the working parameters by more than 10% during 100 years at a service temperature of up to 50°C, during 15–25 years at a temperature of 50–80°C, and for 10 years at temperatures up to 80–150°C. Similar requirements are also imposed on radiation resistance.

In the group of the metallic systems, undergoing structural, phase or structural–phase transformations under thermal and force loading, there are at least two groups which differ both in the nature of formation and behaviour and special features of the structure. In one group, the structural–phase state of the system remains almost unchanged over a wide loading range (these metallic systems are referred to as the materials of quasi-chemical nature). In the other group, the structural–phase state of the system changes under small thermal and force loading and this is consequently accompanied by stress relaxation (these metallic systems are referred to as the relaxation materials). This group contains a wide range of structural–phase states of the system in the vicinity of the stability loss boundary. These states change 'quasi-continuously' under a relatively small change of thermal and force loading. These states are referred to as having low resistance to thermal and force loading. Thus, **the low resistance (or pre-transitional) state of the system is the stage in the vicinity of the structural–phase transformation characterised by anomalies of the structure or properties.** Naturally, the traditionally known structural defects in these specific conditions become already inseparable elements of the structure and interact with each other, and this interaction has a strong effect on the structure and properties of the condensed system. It should be stressed that the density of the structural defects (defects in the traditional sense of the word) is very high and, consequently, they cannot be regarded as isolated and must be studied already as a system of interacting defects in the conditions of the low stability state of the material. This is not a trivial task, especially taking into account the fact that an important role is beginning to be played here not only by the concentration of defects but also their symmetry, the nature of interaction, the depth at which they occur, the type and magnitude of the external effect, and many others. At the background of the state of the condensed system with low resistance to external conditions the role of the interaction of the structural defects becomes very important and sometimes controlling for the structure.

The condensed systems, undergoing phase and structural transformations, are interesting especially due to the fact that their structure in the range of the transitions has the special features characteristic of the transitional processes. It is well-known that the structural–phase transformations are one of the phenomena that are most difficult to describe.

Unfortunately, the books and articles published in recent years pay only little attention to the problem of the thermal and time stability of the structure and properties of amorphous alloys [1–7]. Reviews, concerned with similar problems in nanomaterials, appeared only recently [8–13]. The authors believe that this book is the first attempt for examining this important problem in both the fundamental and applied aspect. As the stability criterion the authors decided to examine the behaviour of the ductility of amorphous alloys because this property is a highly sensitive characteristic of the structural changes which can take place in the amorphous state under the destabilising effects [9]. In the first chapter, the reader can become acquainted with the fundamental characteristics of amorphous alloys which are most important for understanding the structural and physical processes associated with ductility changes. The second chapter is concerned with the main relationships of structural relaxation – the processes leading to the extensive evolution of the structure and properties in the area of the amorphous state under destabilising effects. The third chapter deals with the ΔT-effect discovered by the authors. This effect is associated with the change of the structure and properties during rapid cooling of the amorphous alloys to cryogenic temperatures. Undoubtedly, the fourth chapter is most important because it contains generalised information about the phenomenon of the ductile–brittle transition (temper brittleness) which forms the basis of the approach proposed by the authors to the analysis of the thermal–time stability of the amorphous alloys. The fifth chapter deals with the development of the methods for predicting the stability of the properties, physically justified by examining the phenomenon of the ductile–brittle transition. Finally, chapter 6 presents information on the attempts for the purposeful effect exerted on the structure in order to suppress or prevent the ductile–brittle transition and, consequently, increase the thermal and time stability of the amorphous state.

The authors note that this book does not deal with all the questions. Many aspects remain unclear and disputable. The authors will conclude that their task has been completed if this book will attract the attention of the scientific community to the very interesting and urgent problem of the thermal and time stability of highly non-equilibrium materials with unique properties.

In particular, this book is concerned with the fundamental and, most importantly, physical aspects of the formation, behaviour and special features of the structure and properties of the amorphous state of the condensed systems of advanced materials with low resistance to external influences.

References

1. Glezer A.M., Principles of creating multifunctional constructional materials of a new generation. Usp. Fiz. Nauk. - 2012. - V. 182. - No. 5. - P. 559-566.
2. Glezer A.M., Shurygina N.A., Amorphous nanocrystalline alloys. - Moscow: Fizmatlit, 2013.
3. Melnikova N.V., Egorushkin V.E., Amorphous metals. Structural disorder and kinetic properties. - Tomsk: Publishing house of the NTL, 2003..
4. Filonov M.R., Anikin Y., Levin Yu.B., Theoretical foundations of production of amorphous and nanocrystalline alloys by ultrafast quenching, Moscow: Moscow Institute of Steel and Alloys, 2006.
5. Kherlakh D., et al., Metastable materials from supercooled melts. - Moscow; Izhevsk, Izhevsk Institute of Computer Science, 2010.
6. Belashcheko D.K., Computer simulation of liquids and amorphous materials.- Moscow: Moscow Institute of Steel and Alloys, 2005.
7. Shpak A.P., et al., Magnetism of amorphous and nanocrystalline systems. - Kiev: Akademperiodika, 2003.
8. Andrievsky R.A., Nanostructures in extreme conditions. Usp. Fiz. Nauk. - 2014 - V. 184. - P. 1017-1032.
9. Glezer A.M., et al., The mechanical behavior amorphous alloys. - Novokuznetsk: SibGIU, 2006.
10. Potekaev A.I., et al., Low-stability long-period structures in metallic systems, ed. A.I. Potekaev. - Tomsk: Publishing house of the NTL, 2010.
11. Potekaev A.I., et al. Low-stability long-period structures in metallic systems, ed. A.I. Potekaev. - Tomsk: Publishing house of the NTL, 2012.
12. Potekaev A.I., et al., The effect of point and planar defects on the structural and phase transformations in the pretransition low-stability region of metallic systems, ed. A.I. Potekaev. - Tomsk: Publishing house of the NTL, 2014.
13. Potekaev A.I., et al. Structural features of binary systems with low-stability states, ed.. A.I. Potekaev. - Tomsk: Publishing house of the NTL, 2014.

The main characteristics of amorphous alloys

1.1. Structure

The special features and uniqueness of the physical properties of the amorphous alloys are determined completely by the features of their structural state. From the start of extensive investigations (the middle of the 70s), special attention has been paid to the problems of investigating the structure of amorphous alloys. A number of important results have been obtained using advanced experimental structural methods. Considerable successes have been achieved in the area of computer simulation. However, the problem of the structural state of the amorphous alloys is far from final solution.

Already the first X-ray investigations of the amorphous alloys demonstrated that they do not contain translational symmetry and that their structure is similar to the structure of the liquid. Evidently, the term 'amorphous state', like the term 'crystalline state', assumes the existence of a wide spectrum of different structures, formed in dependence on the production method, chemical composition and subsequent treatment.

It should be noted that the individual models of the structure of the amorphous alloys differ in the construction algorithms (the models based on the random dense packing of the atoms, cluster and polycluster models, the models based on distorted spaces, disclination and dislocation–disclination models), the selection of the atomic interaction potential and the methods of minimisation of energy [1–3]. The models of amorphous substances are constructed using methods of simulation of the atomic systems, like the molecular dynamics method, the statistical relaxation method, the Monte Carlo

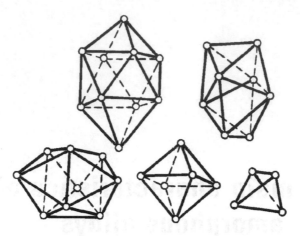

Fig. 1.1. Possible types of Benal's polyhedrons.

method, the methods used for constructing the models on the basis of the available diffraction data on the structure, etc. [2].

At the present time, the proposed series of the structural models of the amorphous alloys [1, 2] can in principle be divided into two large groups [1]: the first group of the models is based on the quasi-liquid description of the structure using the continuous network of the randomly distributed atoms; the second group of the models is based on the quasi-crystalline description of the structure using clusters or crystals, containing a high density of defects of different types.

The atomic structure of the amorphous alloys can be determined by experiments using diffraction investigation methods. The scattering of the X-rays, neutrons and electrons on the amorphous substance makes it possible to determine the general structural factor of the multi-component system $I(k)$ (Fig. 1.1), which corresponds to the sum of the partial structural factors $I_{ij}(k)$ [4]:

$$I(k) = \sum_i \sum_j W_{ij}(k) I_{ij}(k), \qquad (1)$$

where

$$W_{ij}(k) = \frac{c_i c_j f_i(k) f_j(k)}{\langle f(k) \rangle^2}, \qquad (2)$$

c_i and f_i are the atomic concentration and the scattering amplitude of the element i, respectively; $k = (4\pi/\lambda)\sin\theta$ is the length of the diffraction vector; $\langle f \rangle = \Sigma c_i f_i(k)$.

The distribution of the atoms in the amorphous alloy can be determined only by means of the atomic radial distribution function (RDF), and the division of the scattering intensity into the structural factor and the interference function, as was the case in the crystals, is not possible in this case.

The local environment can be analysed by the EXAFS method (examination of the fine structure of the X-ray absorption spectra), using synchrotron radiation [5]. The local fluctuations of the atomic density or concentration are usually analysed by the low-angle scattering method which is sensitive to the variation of the electronic density on the scale of the order of 2 nm.

Because the position of the atoms in the amorphous alloys cannot be determined unambiguously by the diffraction methods, the verification of the structural models is usually carried out by the independent measurement of a number of physical properties (density, heat capacity) and also by investigating the structure by spectroscopic (NMR, NGR) and microscopic (high-resolution electron microscopy, atomic force microscopy) methods.

Most detailed experiments have been carried out into the group of amorphous alloys of the transition metal (TM)–metalloid (M) type whose composition is close to $TM_{80}M_{20}$. For the multi-component systems, which include in fact the most frequently investigated amorphous alloys, the total set of the structural characteristics requires the determination of the partial paired correlation functions. They can be determined most accurately by EXAFS spectroscopy [6].

Although the ensemble of the randomly oriented microcrystals does not have the translational symmetry at large distances, it has been shown [7] that their radial distribution function (RDF) cannot be described by the microcrystalline model, even if the crystals are very small or deformed. In principle, the fact that the microcrystalline model cannot be used reflects the fundamental differences in the nature of the short-range order of the amorphous and crystalline phases. At the same time, there is a number experimental studies in which it is shown that the amorphous state of the majority of amorphous alloys, produced by melt quenching, is of the microcrystalline origin. All these experiments have been carried out by transmission electron microscopy in the regime of formation of the phase contrast. In this case, under specific conditions it is possible to visualise the individual crystallographic planes and even individual atoms.

To describe the structure of the single-component amorphous systems, initial investigations were carried out using the Bernal model [8] which was originally proposed for the description of the structure of simple liquids. The model is based on the random dense packing of rigid spheres ((RDPRS). An important role in the identification of the structure in the framework of the quasi-liquid description is played by the computer simulation methods [1, 2, 9]. However, the methods of successive attachment and collective rearrangement, used in the simulation, did not make it possible to obtain the structure of the randomly close-packed rigid spheres of the same density as that in the experiments. In addition to this, another important problem was the problem of the boundary conditions of the simulated ensemble. In [10] it is proposed to investigate the relaxation of the structure based on the Bernal model assuming the effect of the paired atomic Lennard–Jones potentials which reproduce the radial distribution function for the soft and not for rigid spheres. The model of the random close-packed soft spheres (RCPSS) leads to the considerable improvement of the correspondence of theory and experiment both from the viewpoint of the nature of splitting of the second peak of the radial distribution function and the viewpoint of the density of the amorphous state obtained in the model. The structure in the description by the RDPRS and RCPSS models can be characterised using the Bernal or Voronoi polyhedrons (Figs. 1.1 and 1.2). The Bernal polyhedrons can be used to determine the shape of cavities or potential areas of penetration in the amorphous matrix and contain the atomics spheres at each tip.

The first attempts to solve the simulation problem of the two-component amorphous alloys were made in [11] where was assumed that an alloy of the metal–metalloid type can be simulated using the RCPRS. The 'skeleton' is represented by the metal atoms, and the metalloid atoms occupy the largest cavities. As expected, the number of large-volume cavities in the RCPRS model is not sufficient for distributing 20 at.% of the atoms–metalloids. Nevertheless, the

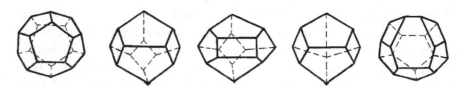

Fig. 1.2. Possible types of Voronoi polyhedrons.

assumption made in [11] resulted in the important qualitative aspect of understanding the structure of amorphous alloys: the coordination cell of the atom of the metalloid consists only of the atoms of the metal and is identical with that observed in the crystalline phases, formed in alloys with a high concentration of the metalloid atoms. For example, in the Ni_3P crystalline compound each phosphorus atom is surrounded by nine nickel atoms, forming a trigonal prism. The identical coordination was observed by experiments in the amorphous alloys of the same composition [12].

Subsequent attempts to construct the models of the structure of the binary amorphous alloys can be divided into two main directions:

1. Computer construction within the framework of the model of the RCPRS of the structure which would then be subjected to relaxation using the appropriate potentials of the paired atomic interactions V_{AA}, V_{BB} and V_{AB}. The final structure should describe accurately the main special features of the general radial distribution function [13]. Although the algorithm of construction of this model for the case of the metal–metalloid alloys assumes the minimisation of the number of the nearest neighbours of the metalloid–metalloid type, the resultant number of these bonds greatly differs from zero.

2. The construction of 'stereochemical' models, proposed by Gaskell [14]. The clusters, consisting of the atom-metalloid and of the metal atoms surrounding it, also form together the coordination cell (for example, in the form of a trigonal prism) (Fig. 1.3). The binary alloys of different composition are regarded as a simple mixture of close-packed areas of the pure metal and areas with the structure of the dense packing of the trigonal prisms in the vicinity

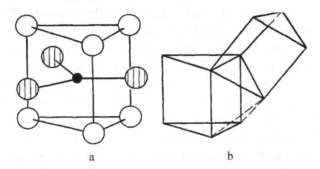

a   b

Fig. 1.3. Trigonal–prismatic element (a) used in the initial configuration of the model by attachment of other such elements (b). The smallest black circle is the atom of the metalloid, open circles are the metal atoms [44]; crosshatched circles are the atoms of the metal situated in the second coordination sphere.

of the atoms–metalloids [15]. The stereochemical model has been studied in the greatest detail in [7, 16] where attention was given gradually to the mechanism and dynamics of the variation of the structure in transition from the liquid to solid amorphous state. Analysis of the suitability of the stereochemical model shows that the model describes accurately the structure of the amorphous alloys of the TM–M type at high concentrations of the metalloid [17]. In all stereochemical models, the knowingly non-periodic packing of the atoms indicates the absence of the periodic potential of the lattice so that the existence in the structure of the effects of the type of vacancies and dislocations is associated with problems, but the concept of inter-cluster boundaries is introduced.

Although the existence of a very strong chemical short-range order in the amorphous alloys of the metal–metalloid has been convincingly confirmed, it is quite difficult to determine its quantitative characteristics. The problem of the topological order in the amorphous alloys is far more complicated than the problem of the compositional (chemical) ordering, because the selection of the ordering parameter is not entirely accurate.

In accordance with the concept proposed by Egami [18], in addition to the topological order it is also important to consider the short-range order of the distortions, associated with the existence in the amorphous matrix of the local geometrical distortions which are in fact structural defects. Together with the topological short-range order, the short-range order of the distortions is regarded as a geometrical short-range order. The topological order in the amorphous alloys is exclusively polytetrahedral. This order is not comparable with the short-range order spatially de-concentrated in the three measurements and existing in the crystals. Nevertheless, it has been shown [19] that in the distorted three-dimensional space (i.e., on the surface of the four-dimensional polytype) the similar perfect polytetrahedral packing becomes possible. To show this structure in the three-dimensional space, it is necessary to introduce defects. These defects are represented, for example, by a number of disclination lines [20]. It has been shown that the ordered series of the dislocation lines describes the Frank–Casper polytetrahedral phase, and the disordered series – the structure of the amorphous state. The problem examined here is very close to the problem of the description of the structure of quasi-crystals experimentally detected in the Al–Mn alloys produced by melt quenching [21]. In fact, the submicroscopic quasicrystals can be used as structural elements when describing the amorphous state.

As already mentioned, the introduction of the disclination considerations is extremely useful when describing the main special features of the structure of the amorphous alloys. In a general case, the defects (linear sources of internal stresses) are referred to as the Volterra dislocations who examined for the first time the methods of producing and describing the stress state for a doubly-connected solid within the framework of the continual theory of elasticity. There are six types of such defects [22].

The concept of the defective lattices assumes the possibility of reaching the amorphous state if the density of the effects is higher than some critical value. It is well-known that to verify the characteristic of any model it is necessary to compare the model with a large number of experimentally obtained data for the structure and, primarily, with the radial distribution functions (RDF) (or paired correlation functions PCF), and also the packing density coefficient ξ. For this purpose, computer simulation of amorphous clusters was carried out in [23] and the PCF and ξ are the functions of the disclination density (or the average distance $\bar{L}$ between them). The initial structures formed by introducing point wedge-shaped disclinations with the Frank vector $\bar{\Omega} = \pm(\pi/3)\bar{i_3}$ to the planar triangular lattice (the positive and negative wedge-shaped disclinations corresponded to 5- and 7-term environments of the central atom).

The disclination theory shows [22, 24] that the dislocations cause Cartan torsion in the crystal without changing its metrics. Therefore, the crystal is investigated in the Euclidean space, and the dislocations are the linear defects of the structure. If the crystal contains disclinations, the latter greatly change its metrics, i.e., the crystal should be regarded as a crystal in the space characterised by the Rieman–Christoffel curvature as a function of the tensor of the disclination density, with the disclinations regarded as the linear defects of its structure. A detailed construction of the models of the structure of the amorphous alloys from the polytypes in the distorted space is used quite widely, mostly for calculating the electronic properties. For example, in [25] it was shown that the dense-packed structures can be produced by imaging the polytype figures from the distorted space to the Euclidean space. Naturally, this imaging is obtained by adding a network of disclination lines transforming the space curvature to zero.

1.2. Structural defects

To understand the role of the defect in a specific process, it is necessary to consider the state of the structure free from defects. Comparison of the state with defects and without them can be carried out in the terms of the topological properties or stress fields. The majority of the defects, typical of crystals, lose their specific features in the amorphous state. Nevertheless, the results of large numbers of experiments carried out to investigate the structure-sensitive properties of the amorphous alloys show that the structural defects can also exist in the amorphous alloys. The deviations in the structure of the amorphous bodies from the low-energy equilibrium state can be described on the basis of the increase of the density of these defects. Taking these considerations into account, the defects cannot be represented by some small disordered region in the amorphous matrix, although small disordered regions in the crystal can be regarded as clusters of elementary defects.

Several attempts have been made to provide a generalised definition of the structural defects in the amorphous solids [26]. On the one hand, a model of the ideal amorphous structure was constructed and defects were then introduced into the model on the basis of the purely geometrical considerations by analogy with what is done in the crystal [27]. This was followed by measuring the resultant displacement which was usually very large. A similar definition of the defects, using the considerations of local deformation, proved to be acceptable only for covalent amorphous solids [27].

Some investigators divide defects in the amorphous alloys into internal and external. The internal defects are typical of the material even after extensive relaxation, and the external defects annihilate during relaxation changes in the structure. Since it is difficult to obtain by experiments only detailed information on different types of defects and their distribution, this is achieved using widely the computer simulation methods which provide data on the defects and their evolution during different external effects [28]. In addition to this, it is evident that these calculations can also help in understanding the amorphous state.

The defects in the amorphous alloys can be divided into point, microscopic elongated and macroscopic. The main point defects, existing in the amorphous matrix, are [27]: broken bonds; irregular bonds; pairs with changed valency; the atoms with a small stress field (quasi-vacancies); the atoms with a large stress field (quasi-implanted atoms). The extended defects include: quasi-vacancy dislocations; quasi-implanted dislocations; the boundaries between two amorphous phases; inter-cluster boundaries. The macroscopic defects include pores, cracks and other macro-imperfections. An important source for the formation of defects in the structure of the amorphous alloys is the free volume determined in fact by a high expansion coefficient of the liquid. Cohen and Turnbull [29] assume that the migration of the atoms in the system, consisting of rigid spheres, becomes possible only in the presence of a cavity which is larger than the critical size. It has been established that the free volume is statistically distributed in the amorphous matrix without any changes free energy during its redistribution within the framework of the same amount. Analysis shows that the energy of formation of the cavity is proportional to T_c and the formation of the cavities of the critical size at temperatures lower than T_c is very difficult. Consequently, the ductility of the system increases.

Therefore, the free volume can be regarded either in the form of cavities of a given size or as a formation continuously distributed in the matrix. In the case of the network models of the structure of amorphous solids (where the continuous, mutually penetrating networks of the atoms are investigated), the positions in which two networks are bonded by the formation of joint atoms are referred to as configuration traps [30]. It is assumed that they form as a result of relaxation which is accompanied by the formation of vacancies in one of the networks. In contrast to the situation examined in the model by Cohen and Turnbull, the distributed volume of the vacancy is stationary because it is always bonded with the already relaxed structure existing around the stationary configuration trap. However, thermal activation may result in the dissociation of the trap and the disappearance of the vacancy which moves to another area forming a new configuration trap. During the lifetime of this vacancy the latter is in principle identical to the cavity in the model examined in [29].

In [31] the authors considered the free volume as areas with reduced density. They are characterised by the size distribution (from the fractions of the atomic radius to the hundreds of nanometres) which depends on the formation conditions, the composition, the heat treatment conditions and a number of other factors. In the existing continuous spectrum of the dimensions of the regions of the free volume different dimensional fractions of these defects should have greatly different activation energies and migration mechanisms. In comparison with other structural models of the defects of the amorphous state, a significant advantage of the theory of the free volume is that it is simple and clear. The model makes it possible to investigate theoretically and sometimes also by experiments the evolution of the regions of the free volume from their nucleation (from the regions of rarefaction in the melt through the transformation of the form and the redistribution in the volume of the amorphous matrix in quenching) to the change of the morphology and the parameters of the size distribution under different external effects (including thermal effects). Regardless of the nature and origin of the free volume, its role in the processes of relaxation, deformation, mass transfer and a number of other processes in supercooled liquids remains controlling.

The examination of the structure of the amorphous alloys by the method of small-angle scattering of the X-rays is in fact the only method capable of providing direct information on the form and size distribution of the defects. Comparative analysis of the experimental data obtained for the basic $Fe_{82.5}B_{17.5}$ alloy and alloys produced by alloying this alloy with 0.01 at.% Sb, Ce or Nb, showed [32] that the introduction of the surface-active elements has a strong effect on the total number of quenched defects and their size distribution.

The scattering heterogeneities in the amorphous alloys are highly polydispersed. Their size range varies from units to several hundreds of nanometres. However, the most important contribution in the statistical size distribution is provided by the submicro-heterogeneities with the size $2R_D \leq 10$ nm. In treatment of the indicatrices by the Guinier method the entire set of observed heterogeneities was formally divided into three conventional fractions corresponding to the finest ($2R_D < 10$ nm), the largest ($2_{RD} > 100$ nm) and intermediate in size (10 nm $< 2R_D < 100$ nm) heterogeneities. The investigated alloys are characterised by the distinctive bimodal size distribution of the scattering heterogeneities, as indicated by the presence of two maxima on the curves of the invariants of the indicatrices (Fig. 1.4) which correspond to the two most probable dimensions <25 nm and lessons ≪150 nm. In all investigated amorphous alloys, the distribution of the heterogeneities with the most probable size (<25 nm) is considerably wider in comparison with the distribution of the larger heterogeneities (the narrower peak on the invariant curve). The ratio of the integral intensities of these maxima depends on microalloying. In the basic $Fe_{82.5}B_{17.5}$ alloy, the bimodal distribution of the heterogeneities is not so distinctive. The distribution of smaller sub-microheterogeneities is characterised by the maximum width with their controlling volume and quantitative contribution to the total concentration of the detected heterogeneities (Fig. 1.4). The most distinctive is the bimodal distribution of the alloy doped with antimony (Fig. 1.5). In the alloy with cerium, the bimodal distribution is more distinctive than in the basic alloy but is less distinctive than in the alloy with antimony. The alloys alloyed with niobium are characterised by the polymodal (instead of bimodal) distribution of the scattering heterogeneities corresponding to the appearance of several maximum on the indicatrix invariant curves (Fig. 1.5). In addition to increasing the polymodality of the size distribution of the heterogeneities, the addition of antimony and cerium results in a large shift of the distribution in the direction of higher values.

The presence or absence of dislocations in the amorphous structure is a relatively disputable question [33]. In particular, in a number of theories it is assumed that the amorphous state is a crystalline state with the dislocation density higher than some critical value (of the order of 10^{14} cm^{-2}). Evidently, the internal dislocations are typical of the amorphous state and cannot disappear during structural relaxation.

Etami et al [34] attempted to describe structural defects in the amorphous alloys from a different and more general viewpoint in the form of sources of internal stresses and a specific type of local atomic symmetry. In this approach, the ideal structure is not required, and all the quantitative characteristics relate to the appropriate equilibrium state which is however not compulsory with the lowest energy of the atomic structure.

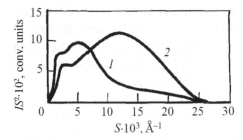 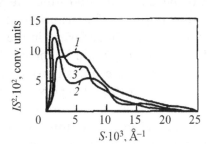

Fig. 1.4. The invariants of the indicatrices of the low-angle X-ray scattering in an Fe–B amorphous alloys: curve 1 – the middle range of a ribbon specimen; curve 2 – the surface.

Fig. 1.5 (right). The invariants of the indicatrices of the low-angle X-ray scattering in the Fe–B (curve 1); Fe–B–Sb (curve 2) and Fe–B–Nb (curve 3) amorphous alloys.

In [35] using the computer model, proposed in [36] for α-iron, it is shown that the distribution spectrum of the local pressure P in the amorphous matrix can be used to determine the extent of structural relaxation: a narrow spectrum corresponds to the well-relaxed state, and a wide spectrum to the non-relaxed state. The change in the distribution of the value P in relaxation may be described on the basis of the redistribution of structural defects determined by the level of the stresses at the atomic level. The recombination of the defects of the p- and n-type explains the change of the density and the main changes of the radial distribution function. The shear defects of the τ-type are not sensitive to structural relaxation.

The correspondence between the theoretical, experimental and simulation results for the structural relaxation of the amorphous alloys makes it possible to determine the number of defects in real materials. For example, it has been established that the annealing of $Fe_{40}Ni_{40}P_{14}B_6$ alloy at 350°C for 30 min results in the recombination of approximately 10% of the defects present in the structure [27].

In [37] A.S. Bakai proposed a polycluster model of the structure of the amorphous state. The polycluster structures are formed by the clusters of atoms, with each cluster characterised by local ordering. The cluster boundaries, investigated in the model, have the form of planar defects and consist of two-dimensional single layers with the imperfect local ordering of the atoms. These boundaries contain a large number of pseudo-vacancies and pseudo-implanted atoms and are responsible for the diffusion and mechanical properties of the polycluster structures.

The presence of the local order in the large part of the volume can be used to determine the structure of the cluster boundaries – the regions of disruption of the local order. In contrast to the intergranular boundaries, in the polycrystalline aggregates described by introducing grain boundary dislocations, the boundaries between the clusters are not simply connected and in fact consist of point defects of different type which merge into complexes.

Within the framework of the polycluster model, the defects are characterised as regions with the highest disordering. They can contain both point and elongated defects. In addition to the regular vacancies and interstitial atoms, they also contain partial point defects (the analogue of the free volume), the cluster boundaries and the one-dimensional defects – the remote analogues of dislocations in crystals – the edges of the boundaries and their sections filled with partial point defects [37]. In addition, the clusters may also contain

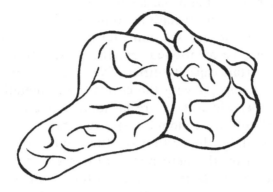

Fig. 1.6. Schematic representation of two connected locally regular clusters.

the areas of localisation with high compression, tension or shear – the analogues of the defects of the n-, p- and τ-type.

Figure 1.6 show schematically two connected (i.e., having the common boundary) locally regular clusters (the sets of regular atoms together with the atoms belonging to their first coordination sphere) whose boundaries are expressed by thick lines. It may be seen that the clusters may have not only external but also internal boundaries, separating the areas of the same cluster. In contrast to the polycrystalline aggregates, the atomic lattices in the polyclusters are generally disordered within the limits of the same whole cluster, although they are characterised by the local order. In addition to this, the clusters, in contrast to the crystals, may have internal boundaries but at the same time the polycrystalline aggregate is a partial case of the polycluster amorphous structure.

1.3. Plastic deformation

The most surprising property of the amorphous alloys is their plastic flow capacity. In fact, plastic deformation is usually regarded as consisting of the nucleation, multiplication and annihilation of the dislocations moving in the solid. However, in the amorphous solid there is no translational symmetry and, consequently, no dislocations in the classic understanding of the nature of this defect. Consequently, the amorphous solid should be absolutely brittle. This situation is characteristic of inorganic glasses, although there are also features of very slight plastic flow in these materials. However, plastic deformation does take place in the amorphous alloys. In this case, it is possible to obtain the anomalously high strength which can be found in the non-crystalline solids if brittle fracture of the solids is prevented at stresses considerably lower than the yield strength. It should be mentioned that the plastic flow capacity of amorphous alloys (this capacity differs from other amorphous solids) is associated undoubtedly with the collectivised metallic nature of the atomic bonds at which there are far more suitable conditions for the processes of collective atomic displacements.

As in the crystals, depending on the degree of reversibility in time, deformation the amorphous alloys can be divided into elastic, inelastic and plastic. Plastic deformation is completely and instantaneously reversible after removing the load, inelastic deformation is completely reversible with time and, finally, plastic deformation is reversible with time after removing the external load. In turn, the plastic deformation in the amorphous alloys may take place by different mechanisms: homogeneously or inhomogeneously. In homogeneous plastic deformation, each element of the solid undergoes plastic shape changes (Fig. 1.7a) because the homogeneously loaded specimen is subjected to homogeneous deformation. In inhomogeneous plastic deformation, the plastic flow is localised in thin discrete shear bands, and the remaining volume of the solid is not deformed (Fig. 1.7b).

Homogeneous deformation is observed in amorphous alloys at high temperatures (slightly below T_c and above this temperature) and at low applied stresses ($<G/100$). Investigations of the homogeneous flow are usually carried out in experiments concerned with creep or stress relaxation. At low applied stresses, the homogeneous flow is

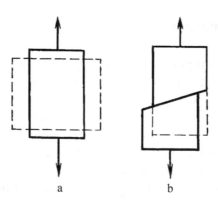

a   b

Fig. 1.7. Diagram explaining homogeneous (a) and inhomogeneous (b) plastic deformation of the specimen.

governed by the law of the flow of the Newtonian liquid: $\dot{\gamma} \approx \tau$. In this case, the shear ductility can be determined using the equation

$$\eta = \tau/\dot{\gamma} \qquad (1.3)$$

where τ is the applied shear stress; $\dot{\gamma}$ is the strain rate.

Homogeneous deformation in the amorphous alloys greatly differs from the viscous flow in the crystals. The main difference is that the steady-state creep stage is not reached in the amorphous alloys, and the strain rate continuously decreases with time. At the same time, the effective ductility, calculated at every moment of time using equation (1.3), continuously increases. It is important to mention the fact that a similar increase of ductility takes place with time according to the linear law in both the freshly quenched and relaxed conditions [38].

A similar behaviour of the amorphous alloys in the homogeneous viscous flow may indicate that the flow reflects in fact the process of structural relaxation and this process is accompanied by a continuous change of the structure. On the one hand, this assumption makes it possible to claim that the atomic mechanism of the viscous flow and of the structural relaxation is identical. On the other and, it is necessary to carry out experiments with the viscous flow of amorphous alloys in the 'pure form' when the changes in the structure do not occur during deformation. A similar viscous flow, referred to as the isoconfiguration flow, can be realised by preliminary annealing at temperatures slightly higher than the test temperature. In this case, the irreversible structural changes can be ignored.

The extent of plastic deformation which can be implemented during the homogeneous flow is in fact not limited. When the temperature, applied stress and the conditions of preliminary heat treatment are correctly selected, the amorphous alloy of the given composition may show the macroscopic deformation of the order of hundreds of percent.

At present, there are several reports according to which some amorphous alloys become superplastic in certain conditions. The superplastic effects in the alloys of the system Ti–Ni–Cu [40] and Pd–Cu–Ni [41] were observed in the tensile test and thermal expansion, respectively. The experimental results show that superplasticity forms in the glass transition temperature range. The experimental results were used to propose a qualitative model of the observed effect which may be described as follows. In the glass transition process

in heating the structural defects are 'defrosted' – the areas in which
the excess free volume is greater than a specific critical value [42].
These defects are responsible for the elementary shear acts, and a
large increase of their mobility in the glass transition range results
in a decrease of the shear ductility in the material and, consequently,
increase of plasticity.

Inhomogeneous flow takes place at low temperatures ($T < 0.8T_c$)
and high applied stresses ($\tau > G/50$). In most cases, inhomogeneous
flow is detected in tensile loading, compressive loading, rolling,
drawing and other methods of deformation of the ribbon, wire and
thick specimens of the amorphous alloys. The inhomogeneous flow
reacts only slightly to the rate of active deformation and demonstrates
the almost complete absence of strain hardening.

The resultant degree of macroscopic deformation is determined
by the number of the shear bands, formed in the amorphous matrix
under external loading. In turn, the susceptibility to the formation of
a large number of shear bands is determined by a number of factors
and, in particular, by the loading method. For example, uniaxial
tensile loading is accompanied by mechanical instability, formation of
a small number of shear bands to catastrophic failure and the general
plastic deformation reaches only 1–2%. At the same time, in rolling
the degree of deformation can be increased to 50–60% [43]. The
highest degree of plastic deformation in the inhomogeneous flow can
evidently be obtained in bending the ribbon specimens. The bending
radius in the limit can be comparable with the thickness of the
ribbon, i.e., may lead to values of 30–40 μm [44]. The characteristic
feature of the inhomogeneous plastic deformation, by which it differs
greatly from homogeneous deformation, is that the inhomogeneous
flow does not increase the degree of order in the amorphous matrix
and in fact reduces this order.

The reports on the conditions of the change of the nature of the
flow of the amorphous alloys are contradicting. For example, it is
assumed that the type of deformation is determined by the absolute
test temperature: at $T < T_g$ (100–150 K) the localised flow must
form, and at higher temperatures the homogeneous flow [45, 46].
It is assumed that the tensile strength depends only slightly on
the temperature and rate of deformation in the case of localised
deformation and strongly in the case of the homogeneous flow. This
assumption corresponds to the claim found in the literature according
to which the type of deformation can be identified on the basis of the
stress–strain diagrams [47]. It is also suggested that the change of the

type of deformation is controlled by the homologous test temperature in the vicinity 0.5–0.7 T_g [48, 49]. Recent studies show that the type of plastic flow and the formation of its relationships are determined by the kinetics of the irreversible structural relaxation [50]. The structural relaxation is defined by both the absolute test temperature and the thermal prior history of the material. If the thermal prior history is such that structural relaxation takes place with the rate maximum possible for the amorphous alloys at the given temperature, the plastic deformation is homogeneous. If the structural relaxation is kinetically 'frozen' (i.e., suppressed by the low test temperature and/or preliminary heat treatment), plastic deformation takes place by the dislocation-like mechanism [50–53]. The change of the plastic flow mechanism is controlled by the transition from one stage of structural relaxation to another, and for the specimens stored after preparation at temperatures not higher than room temperature it takes place in the vicinity of 400 K.

The investigation of the change of the flow mechanisms is a complicated experimental task. The direct structural and metallographic methods of investigation yield only a small amount of information and do not clarify the problem. This may be achieved by using indirect methods which include the measurements of the non-isothermal recovery of the shape of the pre-deformed specimens and the measurements of acoustic emission (AE) during loading. The latter method is more efficient and convincing.

In the case of inhomogeneous plastic deformation shear steps form on the surface of the specimens subjected to preliminary deformation by tensile loading, compressive loading, bending or rolling (Fig. 1.8). These steps correspond to the transfer of the shear bands to the surface. They are usually situated under the angle of 45–55° to the axis of uniaxial tensile loading (or compressive loading) – under the same angle in relation to the rolling direction, and also parallel to the bending axis. The height of the steps above the surface of the specimens reaches 0.1–0.2 μm, and the thickness of the individual shear bands does not exceed (as shown later) 0.05 μm. From this it is clear that the amorphous alloys are characterised by very high local ductility in the inhomogeneous deformation range.

The experiments show that the shear bands are characterised by selective etching [54]; they form in the same areas of the amorphous matrix under repeated loading and at the same time disappear when the deformation of the opposite sign is applied to the specimen [55]. Especially interesting are experiments *in situ* indicating the nature

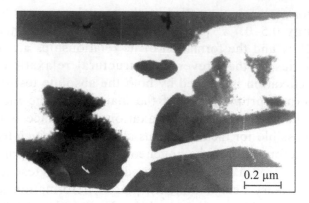

Fig. 1.8. A slip step on the surface of a specimen of $Fe_{80}B_{20}$ amorphous alloys. The image was produced by scanning electron microscopy of the fracture surface.

of nucleation and propagation of the shear bands directly during deformation. Structural studies of this type, carried out by scanning electron microscopy, show [56, 57] that there are three stages of inhomogeneous deformation: the stage of 'homogeneous' deformation in which deformation takes place without recording of the shear bands on the image; the stage in which shear deformation takes place by the formation and propagation of the shear bands and, finally, the final stage in which deformation is localised only in certain bands with the formation of a crack in the 'head' of the bands or in the areas of intersection of the bands. The duration of formation and the propagation rate of the shear bands, measured in [57], were equal to respectively 1.5 μs and $5 \cdot 10^{-5}$ m/s. This was accompanied by the mutual interaction and branching of the individual shear bands and this can be used to explain the small apparent strain hardening observed sometimes on the deformation curves in tensile loading.

In [58] it was shown that the average speed of movement of the shear bands is $10^{-4}-10^{-3}$ m/s, and the slip bands do not interact by intersection but by bypassing an obstacle by the formation of a secondary shear band. It is fully possible that this is associated with the scale factor because the authors of [58] carried out experiments *in situ* by scanning electron microscopy on thin foils of the amorphous alloys.

The deformation generated by the shear band can be described as some macroscopic 'dislocation' or a flat cluster of effective elementary 'dislocations'. Using the model of flat clusters, the authors of [56] simulated the distribution of the shear deformation along the individual band, and the 'dislocations' were represented by

Volterra edge dislocations. The theory describes quite satisfactorily qualitatively the experiment, but only if a certain adjustable parameter is added – the friction force of the amorphous matrix, identical with the Peierls forces in the crystal.

Of special interest is the solution of the problem as to whether there are changes in the structure of the amorphous alloys in the region affected by the inhomogeneous plastic flow. In all likelihood, these changes take place because the presence of a shear band in the amorphous matrix reduces the strength of the matrix that is facilitating further plastic flow in these areas: the material flows far more easily in the areas with local shear and not in the areas which require the nucleation of a new shear band. In the case of ferromagnetic iron- and cobalt-based amorphous alloys cold deformation increases the coercive force by several orders of magnitude as a result of the inhibition of the domain walls by the shear bands [59]. However, both the magnetic and mechanical properties of these materials containing the shear bands can be restored by annealing at temperatures lower than the crystallisation temperature which appears to 'heal' the structure changed in inhomogeneous deformation.

Measurements of the thickness of the shear bands were taken on the $Fe_{65}Ni_{17}P_{18}$ amorphous alloys in different structural states determined by preliminary isothermal annealing at temperatures from 50 to 250°C for 1 hour [60]. Approximately 25 cases identical with those shown in Fig. 1.9 were investigated for each state (including the initial state). The scatter of the values of β was on average up to 50% and this was determined both by the measurement error and by several different starting values of β of different strain bands in the same sample. Regardless of this, there is a clear tendency for a decrease of the value of β with increasing preliminary annealing temperature: from 40±15 nm in the initial state to 20±10 nm after annealing at 250°C (Fig. 1.10).

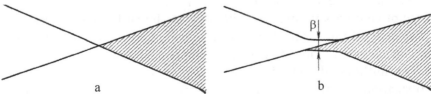

Fig. 1.9. Diagram explaining the determination of the thickness of the shear band.

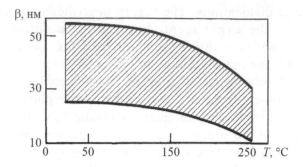

Fig. 1.10. Dependence of the average thickness of the shear band β_{av} on annealing temperature for $Fe_{65}Ni_{17}P_{18}$ alloy.

The experiments indicate that the degree of correlation in the distribution of the atoms which existed in the amorphous matrix prior to the formation is disrupted in the shear bands. Evidently, this results in the sensitivity of the geometry of the shear bands (the thickness) to the state of the amorphous matrix. It is interesting to note that there is a direct analogy between the behaviour of the strain bands in the amorphous alloys and in crystalline solid solutions where the short-range atomic order forms. In the latter case, the moving dislocations also disrupt the correlation in the distribution of the atoms which existed in the solid solution. This creates easier conditions for plastic deformation in the resultant channels and this leads to extensive localisation of the plastic flow in the microvolumes [61].

In [62] the authors carried out detailed analysis of the electron microscope contrast of the shear bands in producing a number of dark field images differing in the position of the aperture diaphragm in the range between the primary beam and the first diffusion maximum. The nature of the changes taking place at the given contrast made the authors of [63] to conclude that the shear bands are characterised by a decrease of the density of the scattering matter. As a result of this, dilation takes place during which the amorphous matrix seems to expand in the shear bands. This phenomenon was also detected in the range of compressive stresses, although as shown by the experiments, the density of the shear bands in the tensile loaded zones of the bent specimens was higher than in the compressed zones.

The authors of [64] developed a mesostructural model of the propagation of shear bands in the amorphous alloys. The conditions for the propagation of the shear bands were derived using the energy approach and the important concept of the fracture mechanics – the J-integral. The stress intensity factor was introduced for the shear bands. This coefficient expresses the measure of the stress concentration at the tip of the band and can be used to describe the conditions of development of the heterogeneous plastic flow for different loading conditions of the specimens of a specific geometry. The resultant relation of the energy balance of the Griffith type:

$$\tau - \tau_0 = \left\{ \left[\left(4\mu / \pi (1-v) \right) \right] \left(\tau_f - \tau_0 \right) u_e / L \right\}^{1/2}, \qquad (1.4)$$

where μ is the shear modulus, v is the Poisson coefficient, τ_f is the maximum stress equal to the local flow stress; τ_0 is the minimum shear stress of the resistance to plastic flow; u_e is the characteristic displacement in the end region of the band, L is the length of the shear band, determines the critical size of the band in relation to the external stresses. The value $J - \tau_0 u^*$ representing the excess energy, associated with the increase of the length of the band during movement of its tip, and the energy dissipated inside the band as a result of overcoming the residual stress of the resistance, should reach the value of the dissipated energy in the end region $\int(\tau - \tau_0) du$ [64]. Formed at the stress raisers, reaching the critical dimensions and propagating, the shear bands become thicker. The thickness of the band can reach the value h, determined by the size of the region r^* in the vicinity of the stress raiser at which the threshold stress of development of localised flow τ_f is reached. The theoretical estimates show that for the stress raisers with the size $L^* = 0.1$–0.5 µm the average thickness of the shear bands reaches $r^* = 10$–50 nm.

1.5. Fracture

As in the crystals, the fracture of the amorphous alloys can be brittle or ductile. In the case of brittle fracture, failure takes place by cleavage without any signs of macroscopic flow at the stress below the yield stress; $\sigma_{fr} < \sigma_y$. In the case of uniaxial tensile loading brittle fracture takes place by breaking of the opposite faces, positioned normal to the tensile loading axis. Ductile fracture in the amorphous alloys takes place either after or simultaneously with the plastic flow process, and the material in this case shows to various degrees features of macroscopic ductility. Here [65]: 1)

fracture takes place on the planes of the maximum cleavage stresses;
2) fracture is constantly associated with one (sometimes with two)
transition from one plane of the maximum cleavage stresses to
another plane; 3) there are two characteristic zones on the fracture
surface: almost smooth cleavage areas and the regions forming the
system of interwoven 'veins'. The thickness of the 'veins' is of the
order of 0.1 µm and they are usually characterised by the ratio of
height to thickness of 2–4, with the exclusion of the points of the
triple junctions.

Figure 1.11 shows typical surfaces of ductile and brittle fracture
produced in the uniaxial tensile loading of $Fe_{82}B_{18}$ amorphous alloys.
The majority of the amorphous alloys are characterised by the classic
pattern of ductile fracture: the fracture energy decreases with a
decrease of the thickness of the ribbon specimens as the conditions
of crack propagation change from the plane stress state to the plane
strain state. For the individual alloys the fracture energy and the
stress for the start of plastic flow are linked by an almost linear
relationship, and the fracture energy increases with a decrease of
the temperature below room temperature. The latter is characteristic
of polymers and is not found in the crystals in which the fracture
energy may decrease to the value of the intrinsic surface energy of
the crack.

As a result of the detailed examination of the parameters of the
plastic zone at the tip of the ductile crack, and also measurements of
the crack opening displacement in dependence on the applied load it

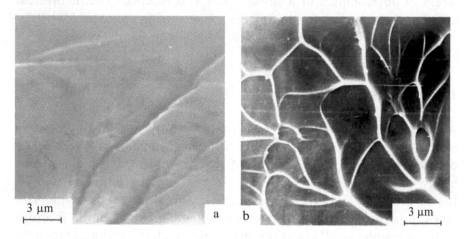

Fig. 1.11. Typical patterns of brittle (a) and ductile (b) fracture in $Fe_{82}B_{18}$ amorphous
alloys. Scanning electron microscopy in reflected electrons.

was found out [65] that there is a direct relationship between the size of the plastic zone, the extent of the crack opening displacement and the applied stress. In addition to this, it was concluded in that the transition from the mixed to plane-stress state at the crack tip in the early stage of plastic deformation and, consequently, a large part of plastic deformation and fracture take place in the plane-stress state. Fracture takes place when one of the following parameters reaches a critical value: the size of the plastic zone, displacement at the crack tip or the stress intensity factor.

The transition from ductile to brittle fracture was studied in detail in [66]. Investigations were carried out on specimens with a V-notch in the $Pd_{78}Cu_6Si_{16}$ alloy after annealing at 350°C for different periods of time. The nature of embrittlement of the amorphous alloys in annealing was investigated in detail in chapter 4 and, therefore, without discussing the phenomenon, attention will be given only to the pattern of the processes taking place. In the initial (quenched) condition and after short-term annealing, the alloy subjected to uniaxial tensile loading failed as a result of gliding after the start of the macroscopic flow. After longer annealing the macroscopic flow was immediately followed by the features of instability of fracture as a result of the intermittent propagation of the cracks. This resulted in the final analysis in the fracture which was preceded by macroscopic flow. Further increase of the annealing time resulted in brittle fracture accompanied by the plastic flow on the surface of the specimens. Finally, the longest annealing resulted in brittle fracture below the elasticity limit without any features of the plastic flow even on the surface of the specimen. Fractographic analysis showed that embrittlement after annealing cannot be attributed only to the increase of the resistance to propagation of the shear bands in the amorphous matrix. The fracture process in the fatigue test starts with the formation in stress concentration areas of the shear bands some of which become crack initiation areas [67]. The plastic zone forms around the fatigue cracks. Crack propagation takes place along the shear bands formed ahead of the crack. This is very similar to the processes taking place in the crystals. The large difference is, however, the fact that the fatigue shear bands in the amorphous alloys are not convex or depressed in the alloy, as in the crystals, but have a single slip step [68]. Analysis of the fatigue fractured showed that there are two stages of crack growth: 1) slow growth leading to the formation of 'terraces' on the surface with fine elements, 2) rapid growth in which the surface consists of coarse striations [68]. The crack growth rate is described by the equation

$$dl/dt = CK_I^n, \tag{1.5}$$

and the size of the plastic zone by the equation:

$$Z_p = AK_I^m, \tag{1.6}$$

where l is the length of the crack; n and m are constants; K_I is the stress intensity factor. This factor determines the growth rate of the fatigue crack through the shear formation process taking place in the vicinity of the tip of the growing crack. In [69] in cyclic loading of the amorphous alloy the results show the formation of cavities ahead of the crack tip, with the crack itself growing in the form of jumps.

The stereoscopic experiments show that the so-called 'veins' on the fracture surface typical of the amorphous alloys are in fact projections on both surfaces. The 'pattern' on the opposite fracture surfaces is similar but not identical. This indicates that the 'veins' formed as a result of the local formation of a neck during fracture. The 'veins' and smooth areas of the fracture surface resemble on the whole the pattern of the surface structure of the crystals formed during fracture in the immediate vicinity of the ductile–brittle transition temperature.

The fact that the 'veins' are areas of exit of strongly localised plastic flow to the fracture surface can be easily identified on the images obtained simultaneously from the fracture surface and the side surface of the deformed ribbon specimens (Fig. 1.12).

Doping with the surface-active elements (surfactants) has a strong effect on the strength of the amorphous alloys. Of considerable interest in this area is the examination of the fractographic special features of the alloys doped with these elements.

The vein-like fracture surface forms only when the fracture takes place in the planes of the maximum cleavage stresses where the plastic deformation processes develop to the highest extent. In the shear bands in inhomogeneous deformation or in the regions of high intensity homogeneous shear in homogeneous deformation the excess free volume is concentrated and the ductility rapidly decreases. The pattern of ductile failure, on the one hand, confirms the plastic flow model based on the dominance of the free volume and, on the other hand, shows how important the ductility parameter is in describing

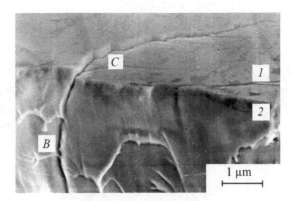

Fig. 1.12. Exit of the shear band to the fracture surface: 1) the side surface; 2) fracture surface; C – slip step; B – the 'vein' (image of two surfaces of the Fe–B alloy obtained by scanning electron microscopy in the reflected electrons mode).

the physical properties and structural processes taking place in the amorphous alloys.

The interesting results for the morphological special features of the micropatterns of failure of the amorphous alloys in microindentation on substrates were obtained in [70, 71]. In the indentation of the amorphous alloys annealed to the temper brittleness temperature ($T < T_{br}$) the indentation was surrounded by the deformation zone formed by heterogeneity deformation reflected in the shear bands originating from the indentation (Fig. 1.13a). In heat treatment at $T > T_{br}$ there were two temperature ranges, and in transition from range to another there was a sharp change of the micropattern of fracture of the amorphous alloys as a result of the local loading. In the first temperature range in the indentation region of metallic alloys there were several radial direct cracks some of which may join with circumferential cracks (Fig. 1.13b). Regardless of the active embrittlement, shear bands can still form. The examined range can be regarded as a transitional range because no cracks form at lower temperatures and shear bands form at higher temperatures. In the second temperature range indentations form mainly prior to crystallisation and represent a system of enclosed squares of cracks which are mutually perpendicular in the areas of the effect of the faces of the Vickers pyramids (Fig. 1.13a). Circumferential cracks appeared in addition to these cracks with increase of the distance from the indentation zone.

It is important to answer the question of the behaviour of the amorphous alloys under the effect of temperature, radiation and

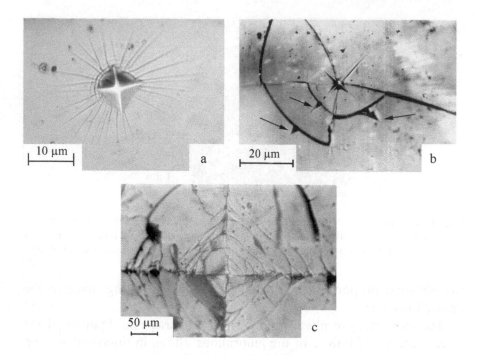

Fig. 1.13. Characteristic indentations on the surface of amorphous alloys of the Co–Fe–Cr–Si system in indentation: at $T < T_{br}$ (a), $T_{br} \leq T \leq 718$ K (b) (the arrows indicate the marks from the face of the indentor separated from the main indentation by circumferential cracks), 718 K $< T \leq 823$ K (c) [144].

external medium. This is important from the viewpoint of the embrittlement susceptibility as a result of specific effects. The problem is reduced used to examining the conditions of development of catastrophic (brittle) fracture in the amorphous alloys under the effect of a number of external factors. The most important parameters of this type include in particular temperature. It is important to distinguish two types of thermal effect on the nature of failure of the amorphous alloys. Firstly, it is the effect of temperature directly during external loading and, secondly, it is the preliminary effect of temperature after which the mechanical tests are carried out as a considerably lower or, on the other hand, considerably higher temperature. The preliminary effect of low temperature (below 0°C) usually has no influence on the mechanical properties and the nature of failure of the amorphous alloys at room temperature. The preliminary effect of high temperatures (starting with T_c and higher) results in a large decrease of ductility and a change of the fracture process.

Transferring to the effect of temperature directly during the mechanical tests, it is important to define two cases: 1) temperature range in which no large temperature change takes place during the test, and 2) the temperature range in which the structure of the amorphous alloy may change during testing. The second temperature range starts at the equicohesion temperature at which the processes of ductile homogeneous plastic flow start as a result of a decrease of the ductility of the material. With increasing test temperature above T_e the classic veiny fracture undergo certain changes: the density of distribution of the 'veins' increases, their form becomes more complicated and, finally, on approach to T_{cr} the fracture becomes absolutely ductile. The fracture surface shows a dimpled relief identical with the relief of the ductile failure of crystals [72]. Low-temperature effects in general features are reduced to the fact that starting with the temperature characteristic for every alloy, fracture takes place by the brittle mechanism without any macroscopic deformation features. A similar ductile–brittle transition can be detected in the $Fe_{70}Ni_8Si_{10}B_{12}$ and $Co_{60}Fe_5Ni_{10}B_{15}$ alloys where the critical temperature of complete embrittlement was approximately 0.2 T_{cr} [72]. The authors of [73] investigated iron-based alloys and obtained the values of the critical temperature in the range 180–230 K in dependence on the specific chemical composition and, in particular, the nickel content. At the same time, there are alloys which do not undergo the ductile–brittle transition at low-temperatures: complete embrittlement of the $Fe_{40}Ni_{38}Mo_4B_{18}$ alloy is not observed up to 77 K [72], and the $Ni_{78}Si8B_{14}$ alloy is not embrittled to 4.2 K [74]. It was assumed that the controlling factor in the absence of the ductile–brittle transition temperature in the amorphous alloy is the high nickel content in relation to the content of metalloid atoms, and in [74] – the presence of disclination elements in the structure of the alloy. It is useful to define two ductile–brittle transition temperatures: T_{br1} – the minimum temperature at which macroscopic deformation is recorded, and $T_{br2} \approx 0.7 T_{cr}$ at which there is a large increase of the ductility as a result of the change of the plastic deformation mechanism from inhomogeneous to homogeneous. In fact, T_{br2} coincides with the equicohesive temperature T_e. It should be however noted that if the point T_{br1} is actually the temperature of the ductile–brittle transition, the point T_{br2} has a less clear physical meaning because the rapid increase of ductility in transition through this point is recorded only in the uniaxial tensile test and, for example, in bending this jump is not detected at all. Nevertheless, the nature of fracture in the ductile homogeneous plastic flow still differs from that in the inhomogeneous flow and, therefore, it may be assumed that the change of the fracture mechanism takes place at the point $T_{br2} = T_e$.

Unfortunately, it is not possible to discuss in detail the problem as to why some amorphous alloys undergo the ductile–brittle transition at low temperatures and other alloys do not. In all likelihood, one can consider the classic consideration of A.F. Ioffe according to which at low temperatures the stress of the start of plastic flow, which depends strongly on temperature in the amorphous alloys, becomes higher than the brittle breaking stress. This leads to the ductile–brittle transition. In the alloys in which the brittle breaking stress is high or the temperature dependence of σ_y is weak, so that the value σ_y does not exceed the brittle breaking stress to the lowest temperatures, the ductile–brittle transition does not take place. Evidently, further studies can provide the actual physical meaning to this phenomenological model.

The amorphous alloys show considerable susceptibility to the embrittling effect of hydrogen in liquid media [75, 76]. It is interesting to compared the effect of hydrogen on the amorphous alloys of the metal–metalloid and metal–metal type. The former (for example, Fe–Ni–B) are characterised by low solubility of hydrogen but considerable embrittlement; the latter (for example, Pd–Zr and Ni–Zr) are characterised by high solubility (up to one hydrogen atom per 1 mole) and the relatively weak embrittlement (Figs. 1.14 and 1.15). It is interesting to note that the alloys in the relaxed state (after annealing) are more sensitive to embrittlement in nitrogen than the quenched alloys. Another interesting special feature is that if the amorphous alloy, saturated with hydrogen, is subjected to vacuum treatment resulting in a rapid decrease of the hydrogen concentration of the alloy, its ductility properties and nature of fracture are completely restored [77].

There is a certain analogy between the effect of hydrogen in liquid metallic media on the brittle fracture susceptibility of the amorphous alloys [75]. In both cases embrittlement takes place leading to a change of the fracture surface: characteristic facets resulting in the 'fish skeleton' appearance of the fracture surface form [77]. This type of faceting was reported in [78] when examining the effect of Hg–In, Hg and Sn–Sb melts on the mechanical properties of several amorphous alloys based on iron. Extensive embrittlement of the material, accompanying faceting, was found.

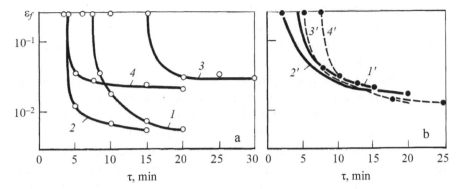

Fig. 1.14. Effect of hydrogen charging time on deformation to fracture of the Fe–Ni–B (a) and Ni–Zr (b) amorphous alloys: curve 1 – $Fe_{39}Ni_{39}B22$ alloy in the quenched condition; curve 2 – $Fe_{39}Ni_{39}B_{22}$ alloy after annealing at 255°C; curve 3 – $Fe_{36.5}Ni_{36.5}B_{27}$ alloy in the quenched condition; curve 4 – $Fe_{36.5}Ni_{36.5}B_{27}$ alloy after annealing at 255°C; curve 1' – $Ni_{67}Zr_{33}$ alloy in the quenched condition; curve 2' – $Ni_{67}Zr_{33}$ alloy after annealing at 250°C; curve 3' – $Ni_{30}Zr_{60}B_{10}$ alloy in the quenched condition; curve 4' – $Ni_{28}Zr_{56}B_{16}$ alloy in the quenched condition.

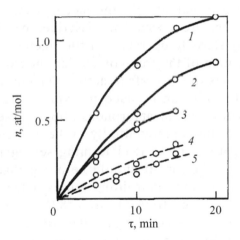

Fig. 1.15. Effect of hydrogen charging time on the hydrogen concentration in the alloys: curve 1 – $Pd_{25}Zr_{75}$ after quenching, curve 2 – $Pd_{25}Zr_{75}$ after annealing at 250°C; curve 3 – $Ni_{67}Zr_{33}$ after annealing at 250°C; curve 4 – $Ni_{30}Zr_{60}B_{10}$ after quenching; curve 5 – $Ni_{28}Zr_{56}B_{16}$ after quenching.

In tensile testing of the $Fe_{40}Ni_{40}P_{14}B_6$ alloy (tensile strain 0.0063) in corrosive media containing acids and chlorides the results published in [79] showed extensive stress corrosion and hydrogen embrittlement. The cracks found in the material propagated in the fan-shaped manner into the opposite side from the nucleation source and were zig-zag-shaped. In [80] cracks of a similar shape were found as a relatively rare case of the fracture of the single-phase ductile metal, determined by brittle fracture at the boundaries of the fragments in the subgrain structure.

The nature of all these phenomena, associated with the special features of fracture of the amorphous materials, can be understood on the basis of the fracture model. The model was constructed on the basis of the controlling role of the melt-quenched free volume and its interaction with the elements decreasing the surface energy. In fact, in any amorphous alloy in the initial condition there is a certain number of the areas of free volume, and its specific characteristics of the size distribution in the amorphous matrix depend on the composition of the alloy and the production conditions. In the absence in the amorphous alloy of the elements decreasing cohesion the fracture process takes place by the mechanism described for the Fe–B amorphous alloy, i.e., shows high micro- and macroductility. However, any treatment of the alloy in the medium resulting in the inflow of the element decreasing cohesion into the volume of the atoms changes qualitatively the mechanisms of micro- and macrofailure. The fracture mechanism, described previously for the Fe–B–Sb(Ce) alloy starts to operate – brittle fracture limited by the 'opening' of the region of the free volume into microcracks and the diffusion of the atoms of the surface active element of the front of the growing crack. This can be used to explain the phenomenon of hydrogen embrittlement of amorphous alloys, observed in [78], because hydrogen diffuses preferentially to the micropores, notches and tips of propagating cracks [81] forming segregations in these areas and reducing cohesion. It should be noted that in diffusion of hydrogen atoms in the amorphous alloys a large number of diffusion atoms is captured from the trap [82] – the region of the free volume, having the size smaller than the critical size for opening into a crack which does not take part in the development of the embrittlement process and even inhibits this process reducing the effective hydrogen content of the matrix. However, the formation of a steep concentration gradient between the surface and the volume during a higher than charging creates suitable conditions for the formation of the adsorbed layer of hydrogen in the majority of regions of the free volume with the size larger than critical. This can be used to explain the considerably greater embrittlement of the amorphous alloys after hydrogen charging in the process of cathodic polarisation in comparison with embrittlement as a result of holding in a 1N solution of HCl [83]. It is interesting to note that the change of the fracture surface in hydrogen charging during cathodic ionisation is identical to the change of the fracture pattern when adding 0.01 at.% Sb (or Cb) to the Fe–B amorphous alloy [84]: there is a transition from the vein-like gap to the fine faceting on the background of the larger specimens. The removal of the free volume from the amorphous matrix may be used to explain also the considerably greater sensitivity of the alloys subjected to the stage of structural relaxation to hydrogen embrittlement.

The effect of hydrogen and deuterium on the structure and properties of the amorphous alloys was studied in a series of works by the authors of [85–88]. The results show a decrease of the elastic properties after hydrogen charging of the amorphous metallic alloys based on cobalt on iron and the subsequent recovery of these processes during holding at 290 5K [86]. The amorphous alloys of the Finemet type saturated with hydrogen over a period of ~10 min showed maximum deformation determined by the losses of elastic properties [87]. On the macroscale this behaviour resembles the variant in which the amorphous material deforms under the effect of intrinsic weight (Newton flow) with increasing temperature as a result of a decrease of ductility. When the duration is greater than 40 min the amorphous alloys start to fracture. It has been noted that the nature of deformation of the amorphous alloys in hydrogen charging and subsequent holding in air strongly depends on the duration of natural ageing (the time passed since preparation). Consequently, it may be assumed that two effects are detected in the hydrogen saturation of the freshly quenched alloys: 1) the reversible loss of the shape; 2) the relaxation of quenching stresses and the relaxation of the free volume. The large number of the observed deformation effects in the alloys based on Si, Ni, Co is explained by the amorphising effects of hydrogen (deuterium) on the matrix. Consequently, a catastrophic decrease of the shear modulus and a change of the topological and compositional short-range order take place [87].

The authors of [88] also reported a large decrease of the shear modulus in hydrogen saturation of $Ti_{50}Ni_{25}Cu_{25}$ alloy in the amorphous state. The addition of hydrogen suppresses or transforms the direct and reversed martensitic transformation B2 $\leftrightarrow$ B19. The results of electron microscopic and X-ray diffraction investigations indicate the increase of the partial fraction of the high-temperature B2-phase in hydrogen saturation of the two-phase $Ti_{50}Ni_{25}Cu_{25}$ alloys. The results obtained in [88] can be used to regard the hydrogen as a structure-forming factor, and its effect on the properties of amorphous alloys is not reduced to the trivial increase of the internal stresses and hydrogen embrittlement.

The effect of high-energy particles on the nature of failure and ductility of the amorphous alloys is highly contradicting. In some studies there was no effect on the mechanical properties, whereas in other studies either the embrittling or plasticising effect was recorded [89]. It is possible that this is associated with the fact that, for example, proton irradiation may result in a number of effects whose influence is often opposite: in implantation of hydrogen, local crystallisation, formation of micro- and macrodefects, local and microscopic heating and, finally, segregation fracture in the amorphous matrix.

1.6. Crystallisation

Up to now, attention has been given to the structural aspect of the mechanical behaviour of amorphous alloys. Undoubtedly, it is also interesting to study the change of the mechanical properties in transition from the amorphous to equilibrium crystalline state. The point is not that one should simulate early the situations associated with the partial transition of the highly non-equilibrium amorphous state to the thermodynamically more stable state under the effect of temperature, time, mechanical treatment, irradiation and other several factors which on its own is also very important. The situation is such that the amorphous–crystalline state may indicate a qualitatively new level of the mechanical properties. The amorphous state is characterised by the low elastic moduli values. Evidently, partial crystallisation increases the value of this characteristic and creates suitable conditions for the more efficient application of amorphous alloys in cases in which the requirements on the materials include not only the high strength and ductility but also sufficiently high values of the Young modulus and shear modulus. The presence of the crystalline phase in the amorphous matrix may also cause undesirable consequences: loss of ductility, structural homogeneity, formation of local stresses, defects, etc. Evidently, many of these factors are determined by the conditions in which the crystalline phase forms because this determines the morphology, phase composition and the number of structural components in the amorphous–crystalline state.

The process of transition from the amorphous to crystalline state can be regarded as the order–disorder type transition. In principle, this may take place either during heating of the amorphous state or in cooling from the melt are the rate close to critical. In the first case, the crystallisation process takes place in the conditions of the constant supply of heat (either at constant or continuously increasing temperature) and under the additional effect of heat generated during crystallisation. Therefore, the structure forms in the system in the majority of cases in a specific stage of heat treatment and consists of two distinctive structural components: amorphous and crystalline [90]. The nature of the structure in this case depends to a certain extent of the heating rate and the subsequent cooling rate, the temperature and the annealing medium.

A completely different morphological type of the structure can be formed in early stages of crystallisation in the conditions of rapid cooling of the melt characterised by a effective heat removal from the solidifying system. Similar amorphous – crystalline formations have been studied only in a small number of cases by the mechanical properties obtained in these cases can be regarded as unique.

Finally, there is another method of formation of the amorphous – crystalline structure when dispersed crystalline particles of refractory compounds (usually, carbides of a refractory metal) are blown using a special nozzle into the 'puddle' of the molten metal. Consequently, the amorphising melt and subsequently the solidified amorphous matrix contains the particles of the crystalline phase uniformly distributed in the volume.

The crystallisation processes taking place as a result of the thermal effect on the amorphous state have been studied in considerable detail.

Detailed discussions have been carried out for the nature of the kinetics of nucleation of crystalline phases. There are a number of disputes regarding the relative role of true nucleation in the amorphous matrix and the athermal growth of the pre-precipitates or frozen-in nuclei. Additional difficulties of interpretation are associated with the measurements of the nucleation rate. The bulk density of the crystallisation centres depending on time can be calculated direct from micrographs. The nucleation rate should be obtained on the basis of measurement of the growth rate and the general transformation kinetics. In the former case, a serious error can be made if the effect of the small cross-section of the thin foils is not efficiently evaluated. In addition to this, in both cases the results obtained in the early stages of crystallisation when the number and dimensions of the crystals are small are incorrect.

In many cases, nucleation takes place homogeneously and the nucleation rate is approximately constant with time at the given temperature. However, the homogeneous nucleation – is only one of the possibilities which is often linked with the heterogeneous nucleation, athermal nucleation and even nucleation, caused by the 'frozen in' crystallisation centres [91]. The surfaces of interfacial boundaries may catalyse the nucleation of centres because the new crystalline phase replaces part of the surface and decreases the total surface energy. However, the heterogeneous nucleation on the surface and interfacial boundaries in amorphous alloys does not take place so extensively as in the transformations in the crystalline solids. The 'frozen-in' centres are sufficiently large to act as crystallisation centres at typical annealing temperatures. They do not always have the irregular structure and become the effective crystallisation centres only after some rearrangement, which requires a certain period of time. At high temperatures, the centres dissolve during annealing.

Systematizing the data on the initial stages of crystallisation, phase transformations taking place during the breakdown of the amorphous state, it may be concluded that the structure, phase composition, the morphology of crystalline phases, formed during the annealing of the amorphous alloys, depends strongly on the production conditions, preliminary heat treatment, and the ratio of the chemical elements of the investigated alloy.

The complicated nature of crystallisation can be determined by calorimetric which determine several distinctive exothermic peaks [92]. The total heat of crystallisation is usually ~40% of the melting point of the alloying; remaining part of enthalpy corresponds to the heat released from the specimen during melt quenching because the heat capacity of the melt is considerably higher.

In many cases, the DSC curve of the glass has two crystallisation peaks [93]. This situation is not unusual: individual weeks, if they are detected, are often close to each other. The double crystallisation peak can be determined by the initial primary crystallisation during which the remaining amorphous matrix changes its composition and then solidifies at a higher temperature with the formation of another phase or, in principle, may be associated with crystallisation without change of the composition followed by crystalline polymorphism transformations (but no such case has been unambiguously observed).

The temperature of the first crystallisation peak depends not only on the heating rate but also on the free volume in the glasses.

In the multistage process, the first peak, corresponding to the crystallisation of the amorphous phase to the metastable crystalline phase, can be smaller than subsequent speaks as a result of the development in further stages of the processes of recrystallisation and phase transitions. The presence in the amorphous alloy of a small amount of dissolved oxygen also results in the transition from the single stage to two-stage crystallisation. Therefore, in the analysis of the information obtained by the DSC it is important to consider the possible role of impurities.

The crystallisation process has been examined on many alloys. However, the main relationships have been established only for the metal–metalloid alloys. The data for the metal–metal alloys are only systematised at the moment. The crystallisation processes in them are slightly more complicated.

All the amorphous alloys of the metal – metalloid type become brittle after the effect of temperature in the temperature range considerably lower than Tg. In this connection, the previously established anomalies of microhardness in the stage of transition from the amorphous or crystalline state are observed on the background of the almost zero ductility of these materials. Therefore, the tensile strength that not have such high values as microhardness because fracture takes place in the elastic range.

The unfavourable effects, associated with embrittlement, can be overcome to a certain degree subjecting the amorphous alloys to heating at a very high rate. In this case, on the one hand, the thermally activated processes forming the basis of the temper brittleness of the amorphous state are suppressed to a certain extent and, on the other hand, the number of crystallisation nuclei, capable of subsequent growth are increase. This leads in the final analysis to the dispersion of the crystallisation products. Experiments carried out on the same $Fe_{70}Cr_{15}B_{15}$ alloy. A ribbon of the amorphous alloy was heated by the direct passage of electric current. The heating rate was 10^3 deg/s resulting in considerable refining of the crystallisation products (Fig. 1.16) and partial retention of the ductility which was typical of the initial amorphous state (Fig. 1.17).

In [91] it was assumed that crystallisation may partially restore the ductility typical of the amorphous state. In [91] appropriate experiments were carried out on amorphous alloys (Fe, Co, Ni)–B. As in any other amorphous alloys showing temper brittleness, in one

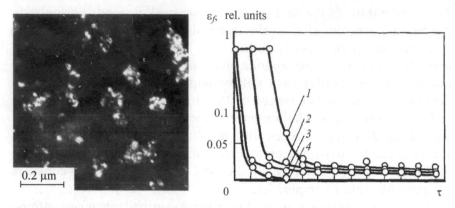

ε_f, rel. units

Fig. 1.16. Structure of the partially crystallised Fe–Cr–B alloy heated at a rate of 10^3 deg/s by direct passage of electric current through the specimen. The dark field in the reflection of α-iron.

Fig. 1.17 (right). Dependence of the ductility ε_f of $Fe_{70}Cr_{15}B_{15}$ alloy on the duration τ of passage of electric current of different intensity through the specimen: curve 1 – 5.4 A; curve 2 – 5.7 A; curve 3 – 6.0 A; curve 4 – 6.3 A.

of these materials, $Fe_{84}B16$ alloy, annealing at 360°C resulted in a rapid decrease of the fracture stress in the uniaxial tensile loading and to the almost complete disappearance of ductility (Fig. 1.18).

At the same time, with the appearance of the first disperse particles of α-iron with increasing annealing time the value σ_p starts to increase again. The maximum value of σ_p of the partially crystallised alloy depends on the microstructure, annealing temperature (Fig. 1.18) and also on the test temperature. An important factor is the one which shows that in the alloy with the optimum content of the α-iron particles brittle fracture is replaced by ductile fracture at the microlevel. Here, as in the quenched amorphous state, there are signs of the characteristic 'vein-like' pattern on the fracture surface. Comparison of Fig. 1.19 (showing the variation of the structural parameters of the amorphous–crystalline state with increase of the annealing time at 360°C in the same $Fe_{84}B_{16}$ alloy) in the data presented in Fig. 1.18, makes it possible to conclude that the maximum value of σ_p corresponds to the bulk density of crystals of α-iron with the maximum particle size of 80 nm.

Longer annealing (as indicated by Fig. 1.18) reduces σ_p and ductility. As assumed by the authors of [94], this effect is associated with the increase of the volume fraction of the particles of the crystalline phase not containing boron. The boron concentration of the amorphous matrix increases and this evidently blocks the

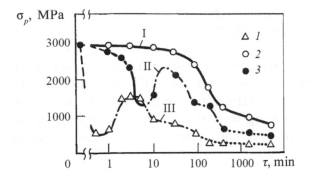

Fig. 1.18. Dependence of the fracture stress σ_p for the $Fe_{84}B_{16}$ alloy on the annealing time at different temperatures, °C: curve 1 – 380; curve 2 – 360; curve 3 – 290; I – amorphous state; II – partially crystalline state; III – completely crystalline state.

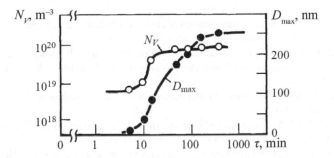

Fig. 1.19. Bulk density N_V of α-iron crystals and the maximum size of crystals D_{max} in dependence on annealing time at 360°C.

migration of the excess free volume. In addition, the local stress fields become larger as a result of the change of the specific volume. The interference of the internal and external tensile stresses may provide a certain contribution to the susceptibility to brittle fracture.

The authors of [94] also determine the phenomenon of plastification (increase of the K_{Ic} parameter in partial crystallisation of the $Co_{75}Fe_4Cr_3Si_{18}$ and $Fe_{58}Ni_{25}B_{17}$ alloys which, according to the authors, is caused by the effective inhibition of the quasi-brittle cracks formed and propagating in the amorphous matrix as a result of external loading. The most effective inhibition, as shown by electron microscopic studies, takes place at the particles of the crystalline phase of the optimum size of 120–140 nm.

Thus, the co-existence of the amorphous and crystalline phases is capable of ensuring some increase of the ductility. As soon as the amorphous matrix is fully crystallised, the ductility appears close to 0. Discussing the reasons for the plastification effect in the stage of introduction of the α-iron crystals, the authors of [95] concluded that the interphase boundaries of the amorphous and crystalline phases may become, under specific conditions, efficient sources of the free volume essential for the efficient course of the plastic deformation processes. It is interesting to note that the identical effect can also be implemented in the mechanical mixture of corundum powder and the appropriate lubricant.

References

1.   Likhachev V.A., Shudegov V.E., Principles of organization of amorphous structures. -SPb .: Publishing house of the St. Petersburg State University, 1999.
2.   Belashchenko D.K., Computer simulation of liquids and amorphous of Greatsubstances. - M .: MISA, 2005. - 408 p.
3.   Poluhin VA Modelling of nanostructures and of precursor states. Ekaterinburg, Ural Branch of Russian Academy of Sciences, 2004. - 208 p.
4.   Wagner K.N.Dzh. Experimental determination of the atomic structure amorphous alloys using scattering methods, Amorphous metalcal alloys. - M .: Metallurgy, 1987. - P. 74-91.
5.   Kekalo IB Atomic structure of amorphous alloys and its evolution. - M .: MISA, 2006. - 340 p.
6.   Carman S.J. EXAFS and XANES studies of metallic glasses, Proc. Fifth Int. Conf. RQM. Elsevier Sci. Publ. - 1985. - V. 1. - P. 427-430.
7.   Gaskell P.H. Local and medium range structure in amorphous alloys, J. Non- Cryst. Sol. - 1985. - V. 75. - No. 2. - P. 329-346.
8.   Bernal J.D. A geometrical approach to the structure of liquids, Nature. - 1959. - V. 183. - No. 4655. - P. 141-147.
       9. Finney JL Simulation of atomic structure, Amorphous metallic alloys. - M .: Metallurgy, 1987. - P. 52-74.
10.   Lancon F., Billard L., Chamberod A. Structural description of a metallic glass model, J. Phys. F: Met. Phys. - 1984. - V. 14. - No. 3. - P. 579-591.
11.   Polk D.E. The structure of glassy metallic alloys, Acta Met. - 1972. - V. 20. - No. 3. - P. 485-491.
12.   Doi K. On a model structure for amorphous solids, J. Non-Cryst. Sol. - 1984. - V. 68. - No. 1. - P. 17-32.
13.   Sadoc J.-F., Mosseri R. Modeling of the structure of glasses, J. Non-Cryst. Sol. - 1984. - V. 61. - P. 487-498.
14.   Gaskell P.H. A new structural model for amorphous transition metals, silicides, borides, phosphorides and carbides, J. Non-Cryst. Solids. - 1979. - V. 32. -No. 1. - P. 207-224.
15.   Gaskell P.H. Is the local structure of amorphous alloys a consequence of medium-range order, Proc. Fifth Int. Conf. RQM. Elsevier Sci. Publ. - 1982. - V. 1. - P. 413-419.
16.   Fujita F.E. On the structure and structural units of amorphous metals, Sci. Repts. Res. Inst. Tohoku Univ. - 1980. - V. A28. - No. 1. - P. 1-7.
17.   Waseda Y., Chen H.S. On the structure of metallic glasses of transition metalmeta loid systems, Sci. Repts. Res. Inst. Tohoku Univ. 1980. - V. A28. - No. 2.

- P. 143-155.

18. Egami T. atomic short-range order in amorphous metal alloys, Amorphous metal alloys. - M .: Metallurgy, 1987. - S. 92-106.

19. Belenky AY Glassy metals, Nature. - 1987. - No. 2. - S. 80-88.

20. Nelson D.R. Structure of amorphous metals, J. Non-Cryst. Sol. - 1984. - V. 61. - No. 3. - P. 475-486.

21. Saito Y., Chen H.S., Mihama K. Icosahedrai quasicrystal in Al-Mn alloy, Appl. Phys. Lett. - 1986. - V. 48. - No. 3. - P. 581-583.

22. De Wit R. Continuum theory of disclination. - M .: Mir, 1977.

23. Likhachev V.A., Hyrum RY Introduction to the theory of disclinations. - L .: LGU, 1975.

24. Zaichenko S.G., Borisov, V.T., About disclination approach to structure of the amorphous state The structure and properties of amorphous alloys. - Izhevsk: Udmurt State University, 1985. - P. 79-83.

25. Sadoc J.F., Periodic networks of disclination lines: application to metal structures. J. Phys. Lett. - 1983. - V. 44. - No. 17. - P. 707-715.

26. Chaudhari P., Defects in amorphous solids, J. Phys. - 1980. - V. 41. - No. 8. - P. 267-271.

27. Popesku M., Defect formation in amorphous structures as revealed by computer simulation, Thin Solid Films. - 1984. - V. 121. - No. 2. - P. 317-347.

28. Kirsanov V.V., Orlov A.N., Computer simulation of atomic configurations of defects in metals, Usp. Fiz. Nauk, - 1984. - V. 142. - No. 2. - S. 219-264.

29. Turnbull D., Cohen M.H., On the free-volume model of the liquid-glass transition, J. Chem. Phys. - 1970. - V. 52. - No. 6. - P. 3038-3041.

30. Krishan K. Structure defects and properties of metallic glasses, J. Non-Cryst. Sol. - 1982. - V. 53. - No. 1-2. - P. 83-104.

31. Glezer A.M., Utevskaya O.L. Parameters of structural relaxation and mechanical cal properties of amorphous alloys, FMM. - 1984. - V. 57. - No. 6. - Pp. 1198-1210.

32. Utevskaya O.L., et al. Anisotropy form of regions of free volume and long-range magnetic order in amorphous alloys. Physics of amorphous alloys. - Izhevsk: Udmurt State University, 1984. - P. 32-36.

33. Steinhardt P.J., Chaudhari P.. Point and line defects in glasses, Phil. Mag. - 1981. - V. A44. - No. 6. - P. 1375-1381.

34. Egami T., Vitek V. Local structural fluctuations and defects in metallic glasses, J. Non-Cryst. Sol. - 1984. - V. 62. - No. 4. - P. 499-510.

35. Egami T., Vitek V., Srolovitz D., Microscopic model of structural relaxation in amorphous alloys, Proc. Fourth Int. Conf. RQM. (Sendai, Japan). - 1981. - V. 1. - P. 517-522.

36. Srolovitz D., Maeda K., Takeuchi S., et al. Local structure and topology of a model amorphous metal, J. Phys. F: Metal Phys. - 1981. - V. 11. - No. 12. - P. 2209-2219.

37. Bakai A.S., Polycluster amorphous structures and their properties. I, Preprint Kharkiv Institute of Physics and Technology. 84-33. - M .: TSNII- Atominform, 1984. - 54 p.

38. Taub A.I., Walter J.L., Scaling the kinetics of flow and relaxation in amorphous alloys, Mater. Sci. and Eng. - 1984. - V. 62. - No. 66. - P. 249-260.

39. Homer C., Eberhardt A. Hot deformation of metallic glass, Scr. Met. - 1980. - V. 14. - No. 12. - P. 1331-1332.

40. Zelensky V.A., et al., Superplasticity of metallic glasses system Ti-Ni-Cu, Physics and chemistry processing machine. Fiz. Khim. Obrab. Mater., 1986. - No. 2. - Pp. 119-121.

41.   Honik V.A., et al., On the relationship between the thermal expansion of amorphous metal alloys with their high-temperature plastic properties, FMM. - 1985. - V. 59. - number 1 - S. 204-205.

42.   Greer A.L. Atomic transport and structural relaxation in metallic glasses, J. Non-Cryst. Sol. - 1984. - V. 61-62. - No. 2. - P. 737-739.

43.   Takayama S. Work-hardening and susceptibility to plastic flow in metallic glasses (Rolling deformation), J. Mater. Sci. - 1981. - V. 16. - No. 9. - P. 2411-2418.

44.   Argon A.S. Inelastic deformation mechanisms in glassy and microcrystalline alloys, Proc. Fifth Int. Conf. RQM. Elsevier Sci. Publ. - 1985. - V. 2. - P. 1325-1335.

45.   Masumoto T., Murata T. Deformation of amorphous metals, J. Mater. Sci. Eng.- 1976. - V. 25. - No. 1. - P. 71-75.

46.   Masumoto T. Mechanical characteristics of amorphous metals, Sci. Rep. Res., Tohoku Univ. - 1977. - V. A26. - P. 246-262.

47.   Inoue A. Bulk amorphous alloys, Amorphous and Nanocrystalline Materials: Preparation, Properties and Applications (Advances in Materials Research) / A. Inoue, K. Hashimoto (eds.). - Berlin, Heidelberg, New York: Springer-Verlag, 2001. - P. 1-51.

48.   Spaepen F. A microscopic mechanism for steady state inhomogeneous flow in metallic glasses, Acta Met. - 1977. - V. 25. - No. 3. - P. 407-415.

49.   Argon A.S. Plastic deformation in metallic glasses, Acta Met. - 1979. - V. 27. - No. 1. - P. 47-58.

50.   Honik VA The role of structural relaxation in the formation of laws plastic flow of metal glasses, Math. Russian Academy of Sciences. Ser. Fiz. - 2001. - T. 65. - No. 10. - S. 1424-1427.

51.   Spivak LV, VA Honik On the nature of the low-temperature anomalies of the inelastic properties of metallic glasses, Zh. - 1997. - V. 67. - No. 10. - S. 35-46.

52.   Vinogradov AY Mikhailov VA, VA Honik Acoustic emission at heterogeneous and homogeneous plastic flow of metal glass, Fiz. Tverd. Tela - 1997. - V. 39. - No. 5. - S. 885-888.

53.   Vinogradov A., Kinetics of structural relaxation and and laws of plastic flow of metal glasses, Fiz. Tverd. Tela. - 1999. - V. 41. - Vol. 12. - P. 2167-2173.

54.   Skakov Yu.A., Finkel M.B., About etching figures in amorphous alloys, Izv. VUZ. Chernaya metallurgiya. - 1986. - No. 9. - PP. 84-88.

55.   Davis L.A. Mechanics of metallic glasses, Prep. Second Int. Conf. RQM. Cambridge University, Cambridge, 1975.

56.   Zielinski P.G., Ast D.G. Slip bands in metallic glasses, Phil. Mag. - 1983. - V. 48A. - No. 5. - P. 811-824.

57.   Neuhäuser H. Rate of shear band formation in metallic glasses, Scr. Met. - 1978. - V. 12. - No. 5. - P. 471-474.

58.   Alekhin V.P., et al. The mechanical properties and structural regularities of deformation and fracture of amorphous Fe-Ni alloys. MiTOM. - 1982. - No. 5. - P. 33-36.

59.   Cronmuller G., Moser N. The magnetic aftereffect and hysteresis loop. Amorphous metal alloys. - Moscow - Metallurgiya, 1987. - P. 338-356.

60.   Glezer, A.M., et al. Electron microscopy study of deformation bands in non-homogeneous plastic flow amorphous alloys, Dokl. AN SSSR - 1985. - V. 283. - No. 1. - S. 106-109.

61.   Skakov Yu.A., Glezer A.M. Ordering and intraphase transformation. Itogi nauki tekhniki. Metalloved. Term. Obrab. - Moscow: VINITI, 1975. - V. 9. - P. 5-72.

62.   Donovan P.E., Stobbs W.M. The structure of shear bands in metallic glasses, Acta Met. - 1981. - V. 29. - No. 6. - P. 1419-1436.

63.   Gerling R., Schimznsky F.P., Wagner R. Influence of the thickness of amorphous

$Fe_{40}Ni_{40}B_{20}$ ribbons on their mechanical properties under neutron-irradiation and thermal annealing, Proc. Fifth Int. Conf. RQM. Elsevier Sci. Publ. - 1985. - V. 2. - P. 1377-1380.

64.   Pozdnyakov V.A. The conditions of formation and development of shear bands in amorphous metal alloys, FMM. - 2002. - V. 94. - No. 5. - p. 26-33.

65.   Inoue A., Takeuchi A. Recent progress in bulk glassy alloys, Mater. Transact. - 2002. - V. 43. - No. 8. - P. 1892-1906.

66.   Peker A., Johnson W.L. A highly processable metallic glass: $Zr_{41.2}Ti_{13.8}$Ni10 $Cu_{12.5}Be_{22.5}$, App. Phys. Lett. - 1993. - V. 63. - No. 17. - P. 2342-2344.

67.   Nishiyama N., Inoue A. Flux treated Pd-Cu-Ni-P amorphous alloy having low critical cooling rate, Mater. Transact. JIM. - 1997. - V. 38. - No. 5. - P. 464-472.

68.   Inoue A. Bulk amorphous and nanocrystalline alloys with high functional properties, Mater. Sci. and Eng. - 2001. - V. A304-306. - P. 1-10.

69.   Egami T. Atomistic mechanism of bulk metallic glass formation, J. Non-Cryst. Sol. - 2003. - V. 317. - P. 30-33.

70.   Fedorov V.A., ET AL. Deformation features and the destruction of tapes of heat-treated metallic glass systems Co-Fe-Cr-Si at microindentation, Izv. RAN. Ser. Fiz. 2005. - V. 69. - No. 9 - C. 1369-1373.

71.   Permyakova I.E. The evolution of the mechanical properties and characteristics of the metallic glass crystallization system Co-Fe-Cr-Si, subjected to thermal treatment: Dissertation - Tula, 2004. - 140 p.

72.   Kim J.-J., Choi Y., Suresh S., Argon A.S. Nanocrystallization during nanoindentation of a bulk amorphous metal alloy at room temperature, Science. - 2002. - V. 295. - P. 654-657.

73.   Zhang T., Inoue A. Ti-based amorphous alloys with a large supercooled liquid region, Mater. Sci. and Eng. - 2001 - V. A304-306. - P. 771-774.

74.   Bengus V.Z., Tabachnikova E.D., Startsev V.I. Mechanical behavior of some metallic glasses at 4.2 to 300 K, Phys. Stat. Sol. - 1984. - V. 81a. - No. 2. - P. K11-K13.

75.   Flores K.M., Suh D., Howell R., et al. Flow and fracture of bulk metallic glass and their composites, Mater. Transact. - 2001. - V. 42. - No. 4. - P. 619-622.

76.   Amiya K., Inoue A. Thermal stability and mechanical properties of Mg-Y-Cu-M (M = Ag, Pd) bulk amorphous alloys, Mater. Transact. JIM. - 2000. - V. 41. - No. 11 - P. 1460-1462.

77.   Filippov VA Investigation of the effect of high density pulse current on the structure and properties of bulk amorphous alloy $Ti_{10}Zr_{40}Cu_{50}$: Dissertation - Moscow, 1993. - 124 p.

78.   Löffler J.F. Bulk metallic glasses, Intermetallics. - 2003. - V. 11. - P. 529-540.

79.   Kakiuchi H., Inoue A., Onuki M., et al. Application of Zr-based bulk glassy alloys to golf clubs, Mater. Transact. JIM. - 2001. - V. 42. - No. 4 - P. 678-681.

80.   Molotilov B.V.,et al. Amorphous alloys. - Moscow: Mashinostroenie, 1986. - 48 p.

81.   Vaillant M.L., Keryvin V., Rouxel T., Kawamura Y. Changes in the mechanical properties of a $Zr_{55}Cu_{30}Al_{10}Ni_{5}$ bulk metallic glass due to heat treatments below 540°C, Scr. Mater. - 2002. - V. 47. - P. 19-23.

82.   Li J.X., Shan G.B., Gao K.W., et al. In situ SEM study of formation and growth of shear bands and microcracks in bulk metallic glasses, Mater. Sci. and Eng. - 2003. - V. A354. - P. 337-343.

83.   Wang Li.M., Wang W.H., Wang R.J., et al. Ultrasonic investigation of $Pd_{39}Ni_{10}Cu_{30}P_{21}$ bulk metallic glass upon crystallization, Appl. Phys. Lett. - 2000. - V. 77. - No. 8. - P. 1147-1149.

84.   Johnson W.L. Bulk glass-forming metallic alloys: science and technology, MRS-

Bull. - 1999. - V. 24. - P. 42-56.

85.   Zhang Y., Zhao D.Q., Wang R.J., Wang W.H. Formation and properties of $Zr_{48}Nb_8Cu_{14}Ni_{12}Be_{18}$ bulk metallic glass, Acta Mater. - 2003. - V. 51 - P. 1971-1979.

86.   Zhang H., Subhash G., Kecskes L.J., Dowding R.J. Mechanical behavior of bulk (ZrHf) TiCuNiAl amorphous alloys, Scr. Mater. - 2003. - V. 49 - P. 447-452.

87.   Inoue A., Zhang W., Zhang T., Kurosaka K. Cu-based bulk glassy alloys with high tensile strength of over 2000 MPa, J. Non-Cryst. Sol. - 2002. - V. 304 - P. 200-209.

88.   Inoue A., Zhang T., Saida J., Matsushita M. Enhancement of strength and ductility in Zr-based bulk amorphous alloys by precipitation of quasicrystalline phase, Mater. Transact. JIM. - 2000. - V. 41. - No. 11 - P. 1511-1520.

89.   Shtremel' M.A. Fracture, Vol. 1. - Moscow: MISiS, 2014. - 670 p.

90.   Skakov Yu.A., Phase transformations during heating and isothermal holding in metallic glasses, Itogi nauki tekhniki. Metalloved. Term. Obrab. - Moscow: VINITI 1987. - V. 21. - P. 53-96.

91.   Koester W., Gerold W. Crystallization of metallic glasses, The metallic glass. Ionic structure, electron transfer and crystallization. Vol. 1. - M .: Mir, 1983. - P. 325-371.

92.   Gupta P.K., Baranta G., Denry I.L. DTA peak shift studies of primary crystallization in glasses, J. Non-Cryst. Sol. - 2003. - V. 317. - P. 254-269.

93.   Kobelev N.P., Soifer YM, Abrosimov GE, et al. High modulus metastable phase in alloys of the Mg-Ni-Y, Fiz. Tverd. Tela - 2001. - V. 43. - Vol. 10. - P. 1735-1738.

94.   Glezer A.M. et al. The plasticizing effect during the transition from amorphous to nanocrystalline, Dokl. RAN. - 2007. - V. 418. - No. 2. - P. 215-217.

95.   Hillenbrand H.G., Hornbogen E., Köster U. Influence of soft crystalline particles on the mechanical properties of (Fe, Co, Ni) -B metallic glasses., Proc. Fourth Int. Conf. RQM. (Sendai, Japan). - 1981. - V. 2. - P. 1369-1372.

2

Structural relaxation

The relaxation process in amorphous alloys has many common features with the extensively investigated process of relaxation in amorphous polymers and oxides but in a number of amorphous alloys it has a considerably stronger effect and influences a larger number of physical and structural parameters [1, 2]. In principle, it is not rational to discuss the properties of amorphous alloys without considering the relaxation parameters, since the measured properties may depend strongly on the extent of structural relaxation of the specific state.

Homogeneous relaxation, also referred to as structural relaxation (SR) takes place uniformly throughout the entire volume of the specimen without influencing its amorphous state. The SR process is accompanied by changes of the short-range order resulting in a small decrease of the degree of non-equilibrium of the glass. The unstable atomic configurations, formed at the moment of amorphisation during quenching, change to the stable configurations by means of small atomic displacements. This results in densening of the amorphous matrix associated with partial annihilation and removal of the excess free volume [3]. It is important to note that the displacement of the atoms during structural relaxation is smaller than the atomic distances and takes place only in localised areas. The magnitude of heat of transformation to the stable phase which can be used as a measure of this non-equilibrium nature changes only slightly in this case.

Structural relaxation is also accompanied by changes of many physical properties of the amorphous alloys: specific heat, density, electrical resistivity, internal friction, elastic constants, hardness, magnetic characteristics (the Curie temperature changes, magnetic anisotropy is induced), corrosion resistance, etc.

A number of models have been proposed for describing the structural relaxation processes and they can be conventionally divided into two groups: 1. The models of the activation energy spectrum or the AES-model. 2. The model proposed by van den Beukel et al.

In the first model it is assumed that the structural relaxation is caused by local atomic rearrangement in the amorphous material, taking place with different relaxation times (activation energies). The fundamentals of the model were described in studies by Primak [4, 5] and later were applied to the relaxation processes in glasses [6–8]. It is assumed that the activation energies of this process are distributed in a continuous smooth spectrum. The rate of variation of the physical properties is proportional to the rate of variation of the density of 'kinetic processes'.

The second model [9–11] is based on the approach to describing the structural relaxation using the short-range classification, proposed by Egami [12]. It is assumed that the first stage of relaxation is accompanied by the compositional (chemical) short-range ordering. This contribution can be efficiently described using the AES model and is a reversible process with the activation energy spectrum from 150 to 250 kJ/mole. The rate of chemical ordering is quite high and after completion of this ordering the topological short-range ordering becomes controlling. Topological relaxation is described by the Spaepen free volume model [13–15] with the unique activation energy of approximately 250 kJ/mole and is an irreversible process. It should be stressed that the main consideration in the Spaepen model is that as a result of the heterogeneous structure of the glass the latter is characterised by the formation of regions with the free volume excessive in relation to the 'ideal structure', i.e., 'relaxation centres'. They may show thermally activated displacements of the atoms which result in the redistribution of the free volume inside the material and also with partial transfer to the free surface.

Because of obvious shortcomings (the direct determination of the magnitude of the excess free volume is not possible, the existence of the unique activation energy of the relaxation processes), the Spaepen free volume model is used only seldom for describing the mechanical behaviour of metallic alloys. The van den Beukel model has also been criticised [16] because it is difficult to imagine the chemical and topological ordering as processes independent of each other and taking place at different times. The models of the activation energy spectrum (AES model) is used most frequently at the present time. In particular, the model of directional structural relaxation, i.e., the relaxation, oriented by external stress, described in [17–20], appears to be promising.

Nevertheless, the change of the majority of the physical properties of the metallic alloys during annealing has been studied in detail and described by both the model of 'topological and chemical ordering, proposed by van den Beukel [21] and by the AES model [7, 17–20, 22, 23].

At the present time, a large number of studies have been concerned with the examination of the behaviour of amorphous alloys during heating, stimulating structural relaxation. The investigations included the determination of the phase transformations, identification of the crystalline phases, determination of their morphology and crystalline structure. The interest in these subjects is caused by the fact that heat treatment may be accompanied by the formation of the amorphous–nanocrystalline and completely nanocrystalline states characterised by the optimum combination of the physical–mechanical properties useful in service.

The authors of [24, 25] observed the non-monotonic temperature dependence of the mechanical properties of thick $Zr_{50}Ti_{16}Cu_{15}Ni_{19}$ amorphous alloys. The rate of decrease of the yield strength with increasing test temperature was higher than the rate of decrease of the tensile strength with increasing temperature. The temperature dependence of the plasticity (ε_{pl}) is conventionally divided into three ranges. At temperatures higher than 575 K (T_g = 590 K) the ductility starts to increase slowly. The transition from the heterogeneous to homogeneous deformation takes place. Subsequently, with increasing T_{an} the value of ε_{pl} rapidly increases, reaching an intermediate maximum of 140% at 685 K. In the range 685–725 K the ductility slightly decreases (to 80%), and the true fracture stress increases at the same time from 250 to 500 MPa. The third temperature range (above 725 K) is characterised by the rapid increase of ductility. The structure was examined to interpret the results [25–27]. The specimens, deformed in the temperature range T_{an} = 293–611 K, remained amorphous. After tensile tests at 658 K the amorphous matrix contains distributed nanocrystals with the size of 10–20 nm. At T_{an} = 685–725 K the structure was nanocrystalline with the grain size of up to 30 nm. The alloy was almost completely solidified and there was a rapid change of the structure of the fracture surface (the 'vein' pattern was almost completely absent). A further increase of test temperature was accompanied by an increase of the size of the nanocrystals to 40–50 nm. Consequently, the experimental results indicate that the mechanical properties in the temperature range 293–658 K are controlled by the amorphous phase, and at higher temperatures by the nanocrystalline structure.

Detailed examination of the mechanical properties in the structure of the amorphous alloys in dependence on temperature was also carried out by the authors of [28]. Experimental material was $Fe_{81.1}C_{13.8}Si_{5.1}$ alloy. This selection was not accidental: it was important to produce in solidification nanocrystalline ferromagnetic alloys characterised by a very low coercive force and high magnetic susceptibility. In addition to this, the search for the compositions of the amorphous alloys based on Fe is determined by the possibility of producing materials characterised by high strength and high ductility at the same time.

Thin bars of the $Fe_{81.1}C_{13.8}Si_{5.1}$ thick amorphous alloy with a diameter of 0.5 mm were characterised by high ductility in bending and a high tensile strength of $\sigma_f \sim 3400$ MPa at 0.4% boron [28]. The fracture surface was characterised by the typical groove pattern on the smooth areas of the surface. In low-temperature heat treatment, with the precipitation of the α-Fe nanoparticles in the amorphous matrix, σ_f reaches 3900 MPa, microhardness HV increases from 9.7 to 11.1 GPa (Fig. 2.1), the Young modulus changes only slightly (from 125 to 128 GPa). Special features of the fracture surface at the annealing temperature of 680 K are similar to the specimens without annealing (Fig. 2.2b), but the area of the smooth sections becomes smaller and the density of the 'veins' increases.

The solidified amorphous $Fe_{81.1}C_{13.8}Si_{5.1}$ alloy has excellent mechanical properties: σ_f in the range 1200 – 2000 MPa, elongation 4–13%. The fracture surface has a typical dimpled appearance (Fig. 2.2c), characteristic of the ductile crystalline alloys in which the crack nucleation sources are distributed uniformly [28]. Consequently, as a result of the formation of quasicrystals in the dispersed state and achieving the optimum distribution of the nanoparticles in the amorphous matrix, the hardening of the matrix becomes evident.

The authors of [29] determined the temperature dependences of the low-frequency internal friction and the shear modulus in the bulk metallic glass of the $Zr_{52.5}Ti_5Cu_{17.9}Ni_{14.6}Al_{10}$ system in the temperature range from room temperature to solidification temperature. The variation of the shear modulus with temperature up to the start of solidification (~700 K) was non-monotonic (Fig. 2.3) [29]: the linear decrease in the range from room temperature to approximately 500 K was almost independent of temperature – the section from

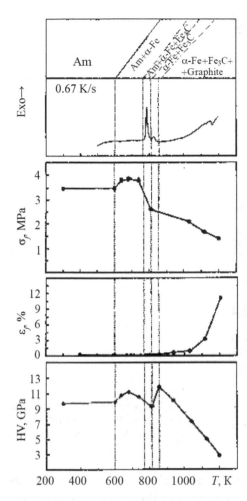

Fig. 2.1. Comparison of the behaviour of the mechanical properties of $Fe8_{1.1}C13.8Si_{5.1}+0.4\%$ B alloy in an annealing with the DSC data and changes in the structure.

500 to 600 K with a small increase in the vicinity of 600 K and the very fast reduction above 600–650 K. The change of the shear modulus G to 400–450 K is reversible, and the magnitude of damping is in the range of the background level. Heating to 500 K and higher is accompanied by the irreversible changes of the increase of G and the decrease of damping, as indicated by the cyclic temperature test.

The authors of [29] defined several aspects in the behaviour of internal friction and shear modulus of the Zr–Ti–Cu–Ni–Al amorphous alloy. The variation of the damping and the shear modulus are qualitatively described by the model of the two-level energy states. The activation energy spectrum of the irreversible and reversible processes of structural relaxation has been evaluated. It

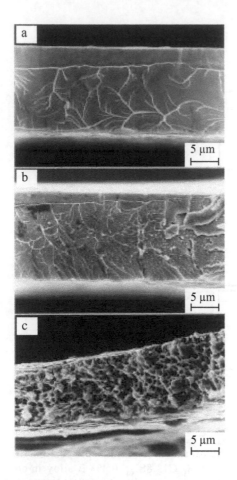

Fig. 2.2. Micrographs of the fracture surface of FC20: (a) initial, (b) T_{an} = 680 K, t = 900 s, (c) T_{an} = 1200 K, t = 900 s.

is assumed that the structure of the amorphous alloys contains the centres which as a result of small changes in the atomic configuration may have two energy states, separated by a barrier.

The observed irreversible changes of the shear modulus are determined by the concentration of the 'non-equilibrium' states. This contribution (non-relaxation) to the elastic moduli is associated with the appearance in the equation for the free energy of the terms proportional to the concentration of non-equilibrium states and the square of the strains formed in this case. In fact, this indicates the non-linear nature of the elastic properties of the material.

To explain the formation of irreversible changes in the damping and the shear modulus, it is assumed that the low-energy states of the centre are localised, i.e., the variation of the orientation (direction of the anisotropy axis) is not accompanied by the change of the centre of gravity and mass transfer. Therefore, up to the

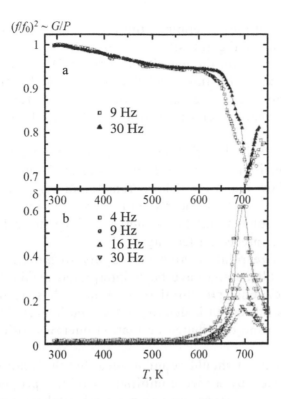

Fig. 2.3. Temperature dependences of the square of the resonance frequency of the torsional tester, proportional in the first approximation to the shear modulus (a), and the damping decrement in the $Ni_{75}Si_8B_{17}$ thick amorphous alloy, containing carbide phase particles (b).

temperatures at which processes with higher activity start to take place, the glass remains in the reversible 'equilibrium' state. The transition to the high-energy state and back may result in the shift of the centre in space. The appearance of the transitions to the state with higher energy in the activation energy spectrum indicates the start of diffusion processes which lead to irreversible structural rearrangements..

The investigation of the kinetics of plastic shape changes of the amorphous alloys is a relatively important task. It is important to mention here the studies [30], concerned with the creep of the $Zr_{52.5}Ti_5Cu_{17.9}Ni_{14.6}Al_{10}$ amorphous alloy. These investigations are essential for both interpreting the deformation mechanism of the amorphous alloys and for explaining the role of the quenching rate – an important factor in the kinetics of formation of the non-crystalline structure.

The authors of [30] determined the following kinetic relationships of creep: 1) after the transition period of $3 \cdot 10^{-3}$ s deformation increases linearly with the logarithm of time; 2) the shear viscosity increases linearly with time, and the angle of inclination $\partial \eta / \partial t$ of the $\eta(t)$ dependences decreases with increasing temperature; 3) preliminary annealing prior to loading results in a large decrease of the creep rate.

The authors of [31] carried out for the first time the measurements of the isothermal relaxation of the stresses of thick amorphous alloys. The results show that after a short transitional period the logarithm of the stresses decreases linearly with increasing logarithm of time and this corresponds to the linear increase of the shear ductility with annealing time, preceding loading.

All the above relationships also apply to metallic glasses in the form of ribbons and have been interpreted on the basis of the model of directional structural relaxation. It has been confirmed that the stress relaxation is determined by the irreversible structural relaxation with the distributed activation energies, oriented by the external effect.

In conclusions, it should be mentioned that the amorphous alloys are characterised by a rare combination of the properties in the structural relaxation stage: unique mechanical characteristics (high strength, hardness, fracture energy, wear resistance), high corrosion resistance, useful magnetically soft properties), high machinability of the material. In addition to this, the production of the amorphous alloys eliminates the problem of producing the required forms of volume semifinished products and components based on them with the isotropic properties in the cross-section. All these advantages cause that these materials can be used as precision materials: magnetically soft materials, fibres or matrices of composite materials; in the manufacture of cutting tools, sporting equipment, mobile telephones, surgery tools, prosthetics, etc.

The application of new technologies of treatment of materials (laser, plasma, chemical–thermal, magnetic pulse, treatment with high density currents, etc) should stimulate the processes of structural relaxation and lead to the required combination of both the surface and volume characteristics of the appropriate component made of the amorphous alloys, with a smooth transition of the properties in the cross-section (gradient materials), and the synthesis of different structures. Special attention should be paid to the method of high plastic strains. In the studies in [32–34] it was shown that the torsion under high quasi-hydrostatic pressure of the amorphous alloys based on iron, nickel, and titanium, results in the self-blocking of the shear bands and delocalisation of the plastic flow and, consequently, in nanocrystallisation and a large increase of the mechanical and magnetic properties.

References

1. Alekhin VP, VA Honik The structure and the physical laws of deformation of amorphous alloys. - Moscow: Metallurgiya - 1992..

2. Zolotukhin I.V., Barmin Yu.V., Stability and relaxation processes in metallic glasses. - Moscow: Metallurgiya - 1991.

3. Betekhtin V.I., et al., Excess free volume and mechanical properties of amorphous alloys, Fiz. Tverd. Tela - 1998. - V. 40. - No. 1. - S. 85-89.

4. Primak W. Kinetics of processes distributed in activation energy, Phys. Review. - 1955. - V. 100. - No. 6. - P. 1677-1689.

5. Primak W. Large temperature range annealing, J. Appl. Phys. - 1960. - V. 81. - No. 9. - P. 1524-1533.

6. Argon A.S., Kuo H.Y. Free energy spectra to inelastic deformation of five metallic glass alloys, J. Non-Cryst. Sol. - 1980. - V. 37. - P. 241-266.

7. Gibbs M.R.J., Evetts J.E., Leake J.A. Activation energy spectra and relaxation in amorphous materials, J. Mater. Sci. - 1983. - V. 18. - No. 1. - P. 278-288.

8. Kruger P., Kempen L., Neuhauser H. Determination of the effective attempt frequency of irreversible structural relaxation processes in amorphous alloys by anisothermal measurements, Phys. Stat. Sol. (A). - 1992. - V. 131. - P. 391-402.

9. Van den Beukel A., Van der Zwaag S. A., Mulder A.L. A semi-quantitative description of the kinetics of structural relaxation in amorphous $Fe_{40}Ni_{40}B_{20}$, Acta Met. - 1984. - V. 32. - No. 11. - P. 1895-1902.

10. Van den Beukel A., Huizer E. On the analysis of structural relaxation in metallic glasses in terms of different models, Scr. Met. - 1985. - V. 19. - No. 11. - P. 1327-1330.

11. Koebrugge G.W., Van der Stel J., Sietsma J., Van den Beukel A. Effect of free volume on the kinetics of chemical short-range ordering in amorphous $Fe_{40}Ni_{40}B_{20}$, J. Non-Cryst. Sol. - 1990. - V. 117/118. - No. 2. - P. 601-604.

12. Egami T., Atomic short-range order in amorphous metal alloys, Amorphous metal alloys. - Moscow: Metallurgiya - 1987. - P. 92-106.

13. Spaepen F. A microscopic mechanism for steady state inhomogeneous flow in metallic glasses, Acta Met. - 1977. - V. 25. - No. 3. - P. 407-415.

14. Taub A.I., Spaepen F. The kinetics of structural relaxation of a metallic glass,m Acta Met. - 1980. - V. 28. - No. 10. - P. 1781-1788.

15. Spaepen F., Taub A.I. Plastic flow and fracture, Amorphous metal alloys. - Moscow: Metallurgiya - 1987. - P. 228-256.

16. Gibbs M.R.J., Sinning H.-R. A critique of the roles of TSRO and CSRO in metallic glasses by application of the activation energy spectrum model to dilatometric data, J. Mater. Sci. - 1985. - V. 20. - No. 7. - P. 2517-2525,

17. Kosilov A.T., Honik V.A. Directional structural relaxation and homogeneous flow of freshly quenched metallic glasses. Izv. RAN Ser. Fiz. - 1993. - V. 57. - No. 11. - P. 192-198.

18.   Kosilov A.T., Khonik V.A., Mikhailov V.A. The kinetics of stress-oriented structural relaxation in metallic glasses, J. Non-Cryst. Sol. - 1995. - V. 192 & 193. - P. 420-423.

19.   Bobrov O.P., Quasi-static and low frequency mechanical relaxation, due to the structural relaxation of metallic glasses: Dissertation -Voronezh, 1996. - 116 p.

20.   Mikhailov V.A. Creep in metallic glasses in an intensive structural relaxation: Dissertation -Voronezh -1998. - 121 p.

21.   Van den Beukel A., Huizer E., Mulder A.L., Van der Zwaag S. Change of viscosity during structural relaxation of amorphous $Fe_{40}Ni40B_{20}$, Acta Met. - 1986. - V. 34. - No. 3. - P. 483-492.

22.   Knuyt G., Stulens H., W. De Ceuninck, Stals L.M. Derivation of an activation energy spectrum for defect processes from isothermal measurements, using a minimization principle and Fourier analysis, Modeling Simul. Mater. Sci. Eng. - 1993. - V. 1. - P. 437-448.

23.   Kasardova A., Ocelik V., Csach K., Miskuf J. Activation energy spectra for stress induced ordering in amorphous materials calculated using Fourier techniques, Phil. Mag. Lett. - 1995. - V. 71. - No. 5. - P. 257-261.

24.   Abrosimova G., Aronin A., Matveev D., et al. The structure and mechanical properties of bulk $Zr_{50}Ti_{16}Cu_{15}Ni_{18}$ metallic glasses, J. Mater. Sci. - 2001. - V. 36. - No. 16. - P. 3933-3937.

25.   Matveev, D.V. Structure and properties of bulk amorphous and nanocrystalline alloys based on Zr and Fe: Dissertation - Chernogolovka - 2004. - 127 p.

26.   Abrosimov G.E.,et al. Formation and nanocrystal structure in a bulk metallic glass $Zr_{50}Ti_{16.5}Cu_{15}Ni_{18}$, Fiz. Tverd. Tela - 2004. - V. 46. - Vol. 12. - P. 2119-2123.

27.   Abrosimova G.E.,et al. Crystalline layer on the surface of Zr-based bulk metallic glasses, J. Non-Cryst. Sol. - 2001. - V. 288. - P. 121-126.

28.   Inoue A., Wang X.M. Bulk amorphous FC20 (Fe-C-Si) alloys with small amounts of B and their crystallized structure and mechanical properties, Acta Mater. - 2000. - V. 48. - P. 1383-1395.

29.   Kobelev N.P., et al., Temperature dependence of the low-frequency internal friction and shear modulus in the bulk amorphous alloy, Fiz. Tverd. Tela - 2003. - V. 45. - Vol. 12. - P. 2124-2130.

30.   Berlev A.E.,et al., Creep of massive metallic glass $Zr_{52.5}Ti_5Cu_{17.9}Ni_{14.6}Al10$, Vestn. Tamb. Gosud. Univ. Estest. Tekh. Nauki. - 2003 - V. 8 - No. 4. - P. 522-524.

31.   Bobrov O.P., et al. Stress Relaxation in massive metallic glass $Zr_{52.5}Ti_5Cu_{17.9}Ni_{14.6}Al_{10}$, Fiz. Tverd. Tela - 2004. - V. 46. - No. 3. - P. 457-460.

32.   Glezer A.M., et al. et al. Self-blocking of shear bands and delocalization of plastic flow in amorphous alloys in sever plastic deformation, Izv. RAN. Ser. Fiz. - 2013. - V. 77. - No. 11. - P.1687-1692.

33.   Glezer A.M., et al. Severe plastic deformation of amorphous alloys. I. Structure and mechanical properties. Izv. RAN. Ser. Fiz. - 2009. - V. 73. - No. 9. - P. 1302-1309.

34.   Glezer A.M., et al. Severe plastic deformation of amorphous alloys. II. Magnetic properties, Izv. RAN. Ser. Fiz. - 2009. - V. 73. - No. 9. - P. 1310-1314.

3

The physical nature of Δ*T*-effect

The amorphous alloys, produced by melt quenching, are, as is well-known, in the metastable equilibrium state. Any destabilising effects (temperature, pressure, deformation, radiation, ultrasound) are capable of causing reversible or irreversible changes in the topological and compositional short-range order and have a corresponding effect on the physical properties of the non-ordered state [1]. The thermal effects are usually reduced to heating of the amorphous alloys above room temperature, but below the crystallisation temperature and, consequently, the processes of structural relaxation, accompanied by the extensive rearrangement of the structure and changes of almost all physical and mechanical properties, take place [2]. At the same time, in a number of investigations [3, 4] the authors reported the possibility of changing the magnetic properties of the amorphous alloys under the effect of cryogenic temperature, for example, low-temperature thermal cycling in liquid nitrogen ($T = 77$ K). It has been assumed that these changes may be caused by the martensite-like phase transition with a large decrease of temperature.

The systematic investigations carried out by the authors of this book show that the phenomenon of the irreversible change of the structure and physical properties of the amorphous alloys after completing low-temperature treatment (LTT) is of the general nature and is typical to some extent of the amorphous alloys [5]. In this chapter, attention is given to the generalisation and theoretical analysis of the previously obtained experimental results in order to formulate the general relationships of the observed phenomenon (low-temperature Δ*T*-effect) and propose its adequate physical model.

The low-temperature Δ*T*-effect has been investigated on amorphous alloys of the $Ni_{81}P_{19}$ binary system (a) for different

superheating temperatures of the melt and on multicomponent
alloys: $Fe_{61}Co_{20}Si_5B_{14}$ (b), $Co_{58}Ni_{10}Fe_5Si_{11}B_{16}$ (c), $Fe_{81}Si_4B_{13}C_2$ (d),
$Fe_{77}Ni_1Si_9B_{13}$ (e) and $Fe_{74.5}Co_{0.75}Nb_{2.25}B_9Si_{13.5}$ (f) [6, 7]. All the alloys
were produced by spinning in the form of ribbons 10–25 mm wide,
with a thickness of 20–25 μm, used for the preparation of specimens
for experimental investigations, including structural, magnetic optics
and thermal studies, and also the measurement of the magnetic and
mechanical properties. The low-temperature treatment of various
duration (from several minutes to 10 hours) was carried out at the
liquid nitrogen boiling point ($T = 77$ K) or the boiling point of liquid
helium ($T = 4.2$ K).

The main mechanical characteristics (Young bending modulus E
and flow stress σ_y) were measured in a dilatometric attachment of the
Dupon thermal analyser by the bending test method [8]. The method
includes the determination of the value E in the elastic section of
loading and the value of σ_y in the elastoplastic section. It should be
noted that the Young bending modulus E coincides with the 'true'
value of E', obtained in the uniaxial tensile test with the accuracy
to the coefficient $(1 - v^2)^{-1}$; the value $E = E'/(1 - v^2)$, where v is
the Poisson coefficient. The measure, reflecting the variation of the
mechanical characteristics of the amorphous alloys was the function
$F = F(E, R, D) = \sigma_y/(2RE)$, where $\sigma_y = f(D)$, $2R$ is the thickness of
the strip, D is the distance between the loading plates. The function
F has a simple physical meaning: its value is equal to the flow stress
of the ribbons made of the amorphous alloy with the unit thickness
and width, and also the unit bending modulus E. Figure 3.1 shows
the dependences of $F(E, R, D)$ on D for the alloys a and b in the
initial state (solid lines) and after low-temperature treatment ($T =
77$ K, $\tau = 2$ h), indicated by the dashed lines. Figure 3.1 shows
clearly the decrease of the flow stress (and also of the yield strength)
of the alloy b, denoted by the digits 1 and 2, by 5%. The flow stress
of the alloy a depends on the melt temperature and, correspondingly,
on the thickness of the ribbon: the thickness of the ribbon of 20.8 μm
correspond to the melt temperature of 1583 K, and the thickness of
the strip of 24.2 μm – the melt temperature of 1393 K. As indicated
by Fig. 3.1, the value of the function F in the first case (curves 3 and
4) decreased by 4.5%, in the second case (curves 5 and 6) by 10%.
In the relative units, the decrease of F for different superheating
temperatures of the melt is ~25%. The maximum increase of the
bending modulus E after low-temperature treatment was obtained for
the alloy a, with the melt temperature of 1583K: 0.5%. The results

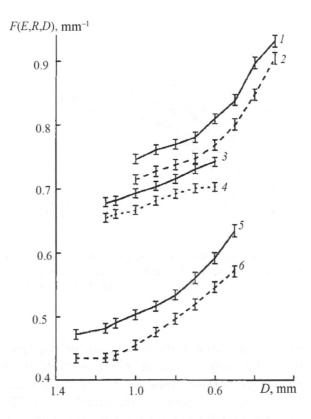

Fig. 3.1. Dependence of the function of the mechanical behaviour of the alloys *a* and *b* on the distance between the loading plates *D*. The solid lines indicate the function $F(E, R, D)$ of the specimens in the initial state, the broken line – after low-temperature treatment ($T = 77$ K, $\tau = 2$ h).

of the mechanical tests of other compositions (*c–e*) also indicated the decrease of the flow stress after low-temperature treatment.

Magneto-optical investigations of the equatorial Kerr effect (EKE) were carried out with the gradual change of temperature during cooling and subsequent heating: $T = 293, 22, 77, 200$ and 290 K. The samples were cooled in a helium cryostat by the blowing method. Experiments were carried out in the spectral magneto-optical equipment, based on the equatorial Kerr effect, in the light energy quantum range from 1 to 3.5 eV. Figure 3.2 shows the temperature dependences of the magneto-optical spectra of the alloy *b*. Similar dependences were also recorded for other amorphous alloys, with the exception of *a* and *f*. For the alloy *e*, the EKE is very small

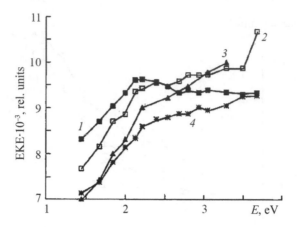

Fig. 3.2. Spectral dependence of the equatorial Kerr effect of alloy *b* on temperature for a gradual change of temperature during low-temperature thermal cycling: curve 1 – T = 290 3K, curve 2 – 22 K, curve 3 – 200 K, curve 4 – 293 K (after completing low-temperature treatment).

($\sim 10^{-4}$) already in the initial state as a result of the low saturation magnetisation of the alloy so that it is not possible to evaluate its changes after low-temperature treatment. The experimental results show that after the low-temperature treatment, the EKE decreases by 10% in comparison with the initial state for both the free (Fig. 3.2) and the contact surfaces of the ribbons of alloy *b*.

The thermal studies were based on comparing the activation energy spectra (relaxation spectra) of the alloys in the initial condition after low-temperature treatment (T = 77 K, the value of τ varied from several seconds to 5 hours). The experiments were carried out by differential scanning calorimetry using the procedure similar to that described in [9]. The variation of the activation energy spectrum as the function of τ (T = 77 K) indicates the gradual decrease of the areas of both the low-temperature and high-temperature parts of the activation energy spectrum (Fig. 3.3). This behaviour of the relaxation spectrum (RS) qualitatively differs from that observed in the case of preliminary annealing at temperatures higher than room temperature when the increase of temperature in heating is accompanied by a decrease (to complete disappearance) of only the low-temperature part of the relaxation spectrum, whereas the high-temperature part remains almost unchanged [9]. The evolution of the activation energy spectrum has another special feature characteristic of the alloys

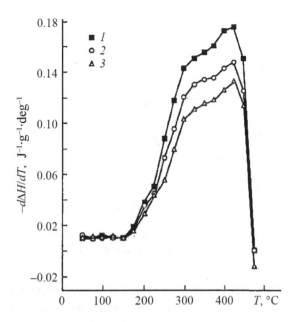

Fig. 3.3. Variation of the activation energy spectrum of the alloy b in dependence on the duration of low-temperature treatment ($T = 77$ K): curve $1 - \tau = 10$ min, curve $2 - \tau = 2$ h, curve $3 - \tau = 5$ h.

based on Fe and Fe–Co: its area S changes almost linearly with the increase of the duration of low-temperature treatment τ. At $\tau > 2$ h the value $S(\tau)$ indicates the weaker dependence on the duration of low-temperature treatment than at $\tau \leq 2$ h.

Special attention was paid to the investigations of the irreversible changes of the magnetic characteristics of the amorphous alloys after low-temperature treatment because these alloys are more sensitive to the low-temperature thermal cycling with large amplitudes of irreversible changes of the main parameters of the hysteresis cycle: coercive force H_c and saturation induction B_T.

Magnetic measurements were carried out by both the static and quasi-static (at the frequency of magnetization reversal of 50 Hz) methods. In the static method, the specimens were in the form of disks with a diameter of 10 mm using a vibration magnetic meter or sections of a ribbon with the size of $100 \times 10 \times 0.02$ mm for the measuring system using a flux meter and a device for measuring the coercive force; in the second method, experiments were carried out with magnetic circuits with the average diameter of 50 mm. Low-

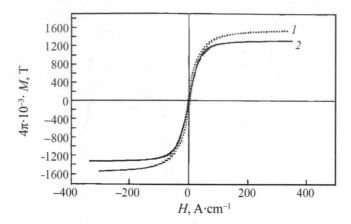

Fig. 3.4. The magnetic hysteresis loop (quasi-static) for the alloy b prior to (curve 2) and after (curve 1) low-temperature treatment (77 K, 2 h). M is magnetisation.

temperature treatment was conducted in liquid nitrogen, the duration of the treatment was varied from several minutes to 10 hours.

The following special features of the low-temperature ΔT-effect, which depend on the composition of the alloy and the parameters of low-temperature treatment, were determined for the magnetic properties:

– for the alloy d based on Fe after low-temperature treatment (T = 77 K, τ = 2 h) B_T increases by 25–30% at an almost constant value of H_c;

– for the alloy b based on Fe–Co after the maximum duration of low-temperature treatment (T = 77 K, τ = 6 h) B_T increases by 7–8% with a simultaneous decrease of H_c by 30% (Fig. 3.4).

– For the alloy e based on Fe–Ni after LTT (T = 77 K, τ = 2 h) H_c decreases by 30% at a small increase of the $B_T \approx 2\%$;

– for the alloy f after low-temperature treatment (T = 77 K, τ = 3–5 h) the value of H_c decreases on average by 30% (the maximum decrease of H_c was 55–60%) at the almost constant value of B_T;

– for all the investigated alloys (with the exception of alloy f, susceptible to nanocrystallisation), high-temperature heat treatment (slightly below the Curie point) does not remove the effect of low-temperature heat treatment: the additive composition of the effects from low-temperature (cryogenic) treatment and from high-temperature treatment takes place (Fig. 3.5).

Of special interest were the investigations of the low-temperature ΔT-effect as a function of the duration of the low-temperature heat treatment for the alloy f susceptible to nanocrystallisation. Analysis

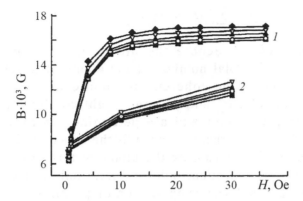

Fig. 3.5. Dependence of the saturation induction of the alloy *b* on the duration of low-temperature heat treatment (T = 77 K), the method of treatment and the strength of the magnetic field: curve 1 – low-temperature heat treatment followed by annealing (T_a = 690 3K, τ = 5 min), curve 2 – only low-temperature heat treatment. Notations: ■ – initial condition, ○– τ = 1 h, ▲ – τ = 2 h, △ – τ = 3 h, ♦ – τ = 6 h.

of the results shows that in addition to the previously mentioned special features, this alloy also showed a number of other special features:

– the significant effect of the low-temperature heat treatment on the irreversible changes of the magnetic characteristics was observed only at the optimum duration of the heat treatment (3–5 h);

– a large decrease of H_c was recorded only for the specimens which could not contain the nanocrystalline phase (or nuclei of this phase) in the initial state; in the specimens subjected to nanocrystallisation (even early stages of nanocrystallisation) already during melt quenching, the value of H_c decreased only slightly (by no more than 4–5%) or did not change at all;

– the decrease of the value H_c for the given treatment method was obtained only after the first two consecutive low-temperature cycles; the subsequent cycles did not result in any change of H_c.

The experimental investigations of the irreversible structural changes, caused by low-temperature heat treatment, were carried out by neutron diffraction and Mossbauer spectroscopy.

Neutron diffraction studies. The intensity $I_{obs}(Q)$, measured in the neutron diffraction experiments with the amorphous alloys, in the static approximation is determined as $I_{obs}(Q) = AS(Q)$, where A is the interaction constant, $S(Q)$ is the static structural factor. For multicomponent amorphous alloys, $S(Q)$ is the weight sum of the partial functions having the form

$$\sigma S(Q) = \sum_{i,j}^{n} \alpha_{i,j} S_{i,j}(Q) + \sum_{i}^{n} \beta_{i} S_{i}^{inc}(Q), \qquad (3.1)$$

where the indexes i, j denote the atoms of the type i and j, respectively; n is the total number of the atoms of different sort; $S_{i,j}(Q)$ and $S_{j}^{inc}(Q)$ is the coherent and non-coherent scattering function, respectively; $\alpha_{i,j}$ and $\beta_{i,j}$ are weight coefficients for the partial scattering functions and are determined as follows $\alpha_{i,j} = 4\pi c_i c_j b_i b_j$ and $\beta_i = c_i \sigma_i^{inc}$ where c_i and b_i is the concentration and the maximum length of scattering by the atoms of this sort; σ_i^{inc} is the elastic scattering cross-section, index i relates only to the atoms of the sort i; σ is the total scattering section of the alloy:

$$\sigma = \sigma_{coh} + \sigma_{inc} = 4\pi \sum_{i,j}^{n} c_i c_j b_i b_j + \sum_{i}^{n} c_i \sigma_i^{inc}, \qquad (3.2)$$

where σ_i and bi are linked by the relationship $\sigma_i = 4\pi b_i^2$. For the $Fe_{61}Co_{20}B_{14}Si_3$ alloy, the bonds Fe–Fe and Fe–Co provide a contribution greater than 80% to the total structural factor $S(Q)$ because the contributions from the bonds of the Fe–B and Fe–Si atoms equal 5% each. The contribution from other bonds are negligible.

To determine the changes of the short-range order in the $Fe_{61}Co_{20}B_{14}Si_5$ alloy, caused by low-temperature heat treatment with the parameters ($T = 77$ K, $\tau = 2$ h), where τ is the holding time of the amorphous alloy in liquid nitrogen, experiments were carried out to compile the structural factors on the freshly quenched specimens of the alloy subjected to low-temperature heat treatment. The neutron diffraction experiments were carried out in an LAD diffractometer with the pulsed neutron source in the Rutherford Laboratory (England). The intensity of scattering of the neutron beam incident on the specimen was recorded simultaneously by 14 blocks of detectors with pulse counters. The incident beam contained neutrons with a large number of wavelengths and, therefore, to obtain the total structural factor $S(Q)$ the data from all 14 detectors were added up. The isotropy of the specimens was obtained by filling a vanadium container with an outer diameter of 8 mm with ribbon specimens of the alloy with the area of approximately 10 mm². The experiments were carried out at room temperature

The radial distribution function (RDF) $g(r)$ of the freshly quenched specimens of the $Fe_{61}Co_{20}B_{14}Si_5$ alloy is shown in Fig. 3.6. The function $g(r)$ of the specimens after low-temperature heat treatment

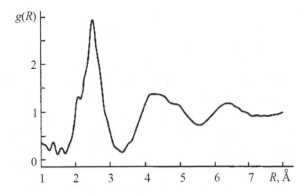

Fig. 3.6. The radial distribution function $g(r)$ of the freshly quenched amorphous alloy $Fe_{61}Co_{20}B_{14}Si_5$ with the parameters of low-temperature heat treatment ($T = 77$ K, $\tau = 2$ h).

($T = 77$ K, $\tau = 2$ h) is similar to that shown in Fig. 3.6, but the scale of the figure does not make it possible to see the changes from $g(r)$ obtained for the freshly quenched specimens.

It should be noted that the radial distribution function (RDF) is connected with the structural factor $S(Q)$ by the relationship

$$g(r) - 1 + \left(2\pi^2\rho_0 r\right) \int_0^\infty [S(Q) - 1] Q \sin(Qr) dr,$$

where ρ_0 is the density of the amorphous alloy. The data indicate that both specimens are completely amorphous. This is confirmed by the absence of diffraction peaks characterising crystalline inclusions at the maximum scattering angles with the resolution $\Delta Q/Q \leq 0.5\%$. The difference between the RDF of the freshly quenched specimens and the specimens subjected to low-temperature heat treatment is in the positions of the maxima of $g(r)$. In order to determine these displacements accurately, the data for each block of the detectors were processed separately and approximation was carried out using the Lorenz function. Consequently, it could be concluded that the displacement of the first maximum of the structural factor $S(Q)$ for the specimen after low-temperature heat treatment is in the range of the experimental error, the displacement of the second maximum in the direction of increasing values of Q is 0.06 ± 0.02; the third maximum 0.12 ± 0.03 nm^{-1}.

As already mentioned, the main contribution to the neutron scattering intensities is provided by the pairs of the Fe–Fe and Fe–Co atoms. This is in agreement with the position of the first maximum of $S(Q)$ observed for $r = 0.251$ nm because this value corresponds to the mean weighted composition with the radii of the atoms of Fe and Co equal to 0.127 and 0.125 nm. For small values of r the left branch of the first maximum of $g(r)$ shows a distinctive preliminary peak corresponding to the Fe–B and Co–B bounds. Similar neutron diffraction experiments were carried out in the time-of-flight diffractometer DN-2 of the IBR-2 reactor at the Joint Institute for Nuclear Research (Dubna) for the $Fe_{61}Co_{20}B_{14}Si_5$ alloy with other parameters of low-temperature heat treatment ($T = 77$ K, $\tau = 3$ s) and the amorphous alloys susceptible to the nanocrystallisation, $Fe_{78}Cu_1Nb_4Si_{13}B_4$ alloy with the low-temperature heat treatment parameters ($T = 77$ K and $\tau = 3.5$ h). The increase of the duration of low-temperature heat treatment for the $Fe_{78}Cu_1Nb_4Si_{13}B_4$ alloy increases the displacement of the second and third maximum of this alloy and equalled 0.08 ± 0.02 and 0.16 ± 0.04 nm^{-1}, respectively, and the displacement of the first maximum was within the experimental error range. In the $Fe_{78}Cu_1Nb_4Si_{13}B_4$ alloy, low-temperature heat treatment resulted in considerably larger displacements of the maximum in the direction of increasing values of Q: displacement of the preliminary peak was 0.03 ± 0.001; the first, second and third maximum 0.098 ± 0.03; 0.17 ± 0.06; 0.28 ± 0.05 nm^{-1}, respectively [10].

   The Mossbauer spectra were measured in the initial condition and after low-temperature heat treatment ($T = 77$ K, $\tau = 2$ h). The source of γ-radiation was 57 Co with the activity of 50 mCi. The comparison of the effective magnetic fields H_{eff} after low-temperature heat treatment for the alloy e indicates that the increase after low-temperature heat treatment was 2%. Analysis of the distribution function of H_{eff} shows that the change of the short-range order after low-temperature thermal cycling may result in the enrichment of the nearest neighbourhood of the Fe atoms by the atoms of Si and B [11].

   The low-angle neutron scattering. To determine the correlation between the changes of the mechanical characteristics and the structure of the amorphous alloys caused by low-temperature heat treatment, experiments were carried out using the $Ni81P_{19}$ binary alloy because the low-angle neutron scattering (LANS) provides the most accurate and reliable results especially for the binary alloys of the transition metal–metalloid type. The LANS experiments were carried out in a LAD diffractometer with a pulsed neutron source at the Rutherford Laboratory (England). The neutron pulses were in the range $0.1 \leq Q \leq 2.4$ nm^{-1}, the specimens were bombarded

with electrons with the wavelengths in the range from 0.22 to
1 nm. The time-of-flight procedure corresponded to the wavelengths
of the neutrons located in the real space in the range from 3 to
60 nm. All measurements were made taking the background scattering
into account and reduced to the absolute units of the scattering cross
sections using the standard correction procedure. Specimens of the
$Ni_{81}P_{19}$ alloy were placed in the form of a stack of parallel ribbon
specimens secured with a special clamp. They were placed in a
vanadium container. Before the measurements it was established
that the windows of the container do not provide any significant
contribution to the LANS. The width of the neutron beam was
smaller than the width of the stack of the ribbons; the thickness
of the latter was 1.1–1.3 mm. The recording time of the scattering
curve was usually 1 h.

The LANS was measured for specimens of the $Ni_{81}P_{19}$ alloy
produced with three different methods of heat treatment of the
melt prior to quenching. The ribbon from which the samples were
produced was marked with number 1 and was produced by melt
quenching, with the melt temperature being 1383 K. Number
2 denotes the samples produced from the ribbon made by melt
quenching at a temperature of 1593 K. The heat treatment of the
melt used for producing the ribbon for samples No. 3 consisted of
heating from 1383 to 1593 K, holding at the latter temperature for 5
min and followed by cooling to 1383 K and then by melt spinning.
According to the phase diagram, the melting point of the eutectic
with the composition $Ni_{81}P_{18}$ was 1136 K. All three types of samples
were studied in detail by high resolution neutron diffraction which
was described above.

The experimental results confirm that the ribbon for the samples
1 was completely amorphous. The absence of Bragg peaks on the
diffraction pattern taking the experimental resolution into account
indicates that this ribbon is equivalent to the ribbon containing less
than 0.05 volume concentration of the crystalline phase. The ribbon
for the samples 3 contained a relatively large number of crystallites
(of the order of 1 vol.%) whereas the ribbon for the samples 2
contained no more than 0.2 vol.% of the crystallites. After quenching
the ribbons were held for several hours at room temperature. During
this period no processes associated with structural relaxation took
place (according to differential scanning calorimetry data). The
low-temperature heat treatment of the ribbons for the samples 1–3

was carried out in liquid nitrogen ($T = 77$ K) for 2 h by the method described in [9].

The macroscopic differential scattering cross section (MDSC) which determines the scattering intensity of LANS for diluted solid solutions and isotropic systems, containing non-interacting, randomly distributed particles, has the following general form [8]:

$$[d\Sigma / d\Omega](Q) = V_T^{-1} \sum_{i=1}^{S} u_i \int_{a_i} W(a_i) V_{P_i}^2 |f(Q,a_i)|^2 da_i, \qquad (3.3)$$

where Q is the scattering vector, $f(Q,a_i)$ is the value taking into account the form of the particles, a_i is the characteristic of the linear size of the scattering object, $W(a_i)$ is the distribution function of a_i, V_{P_i} is the volume of the scattering particles with the contrast u_i, determined as the square of the difference between the densities of the scattering lengths from the particle and the medium surrounding it, V_T is the volume of the sample. Summation is carried out for all scattering objects representing different phases. Equation (3.3) shows that in the LANS experiments the intensity of the measured signal is proportional to contrast u_i and not to the scattering length of the individual nucleus, similar to standard neutron diffraction experiments. In the present case the $Ni_{81}P_{19}$ alloy in dependence on the type of heat treatment of the melt contains crystalline particles of the same composition and, therefore, equation (3.3) can be simplified [11]:

$$[d\Sigma / d\Omega](Q) = \left(V_p^2 N_p u_p^2 |F_p(Q)|^2 / N \right), \qquad (3.4)$$

where V_p is the volume of particles, N_p is the number of particles with contrast u_p. It is shown in [12] that at low values of the product $Q \times a$, where a is the linear size of the particle, the value $|Fp(Q)|^2$ can be approximated in the form $|F_p(Q)| \approx \exp\left[-Q^2 R_G^2 / 3\right]$ for the particles of any shape. R_G is the radius of rotation, for spherical particles with a radius R it is equal to $R_G = 3.5^{1/2}R$. This approximation is a good one when the product $Q \times R_G$ is lower than unity. When the product $Q \times a$, where a is the smallest linear size, the approximation (3.4), as shown in [11], can be used to obtain information on the investigated object using the measured curves of low-angle neutron scattering. For homogeneous particles with sharp boundaries, the equation (3.3) can be simplified

$$[d\Sigma / d\Omega](Q) = k_p / Q^4 = \left(2\pi n_p S / \left(V_T Q^4\right)\right), \qquad (3.5)$$

where k_p is a constant, proportional to the total surface area of crystalline particles. The volume concentration of the latter can be obtained by integration of equation (3.3), because the integral of the scattering cross-section, the so-called invariant, is linked with c_p by the relationship [11]

$$\int Q^2 [d\Sigma / d\Omega](Q) = 2\pi^2 \left(1 - c_p\right) c_p u_p^2, \qquad (3.6)$$

and therefore for large Q the macroscopic differential scattering cross section (MDSC) can be written in the form

$$[d\Sigma / d\Omega](Q) = C_1 + C_2 / Q^{C_3}. \qquad (3.7)$$

If the approximation (3.3) is accurate [10], the constant C_2 is equal to the constant k_p, and the constant C_3 is equal to 4. The value of C_3, determined for specimens, is equal to 4.00 ± 0.06 and, therefore, the approximation (3.3) should be regarded as valid.

The dependences of MDSC, measured for all types of the freshly quenched specimens of $Ni_{81}P_{19}$ alloy, are presented in Fig. 3.7. There is no background scattering and no instrument errors and they have been reduced to the absolute values of the scattering cross-section. The dependences of MDSC for the freshly quenched specimens, denoted by the digits 1 and 2, are similar to each other and at the same time, the dependence of the MDSC for the specimens 3 is greatly different indicating the considerably larger scattering cross-section. The specimens 3 contain approximately 1 vol.% of the crystalline phase, and this explains both the higher value of MDSC and the local maximum on its dependence in the range $Q \approx 0.3$ nm^{-1}. This value of Q corresponds in the real space to the crystals with the characteristic linear size of the order of 20 nm. The main conclusion, resultant from the dependence of the MDSC (Fig. 3.7), is that all the freshly quenched specimens of the alloy contain both the amorphous and crystalline phases. In the specimens 3 the crystalline phase corresponds to the particles with the linear size of ~20 nm.

The effect of thermal cycling in liquid nitrogen of the ribbons of $Ni_{81}P_{19}$ alloy is clearly demonstrated by Fig. 3.8 which shows the differences in the MDSC of the freshly quenched specimens and the

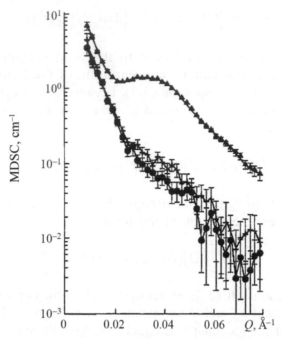

Fig. 3.7. Macroscopic differential scattering cross sections (MDSC) of the $Ni_{81}P_{19}$ amorphous alloy. Notations: specimens 1 – black circles; specimens 2 – crosses; specimens 3 – triangles.

specimens subjected to low-temperature heat treatment. On the basis of the comparison of Figs. 3.8a and c it may be concluded that the size of the particles of the crystalline phase after low-temperature heat treatment changes irreversibly in the direction of low values at the initial volume concentration of 0.05 and <0.2 vol.% in the freshly quenched specimens, respectively. In the specimens 3 there is no decrease of the concentration in the linear dimensions of the crystalline particles after low-temperature heat treatment which is confirmed by the almost complete agreement of the MDSC in the freshly quenched specimens and the specimens subjected to low-temperature thermal cycling (Fig. 3.8b). The comparison of the structural studies with the results of mechanical tests of the $Ni_{81}P_{19}$ amorphous alloy is given below.

On the basis of the experimental results it may be assumed that the nature of the observed low-temperature ΔT-effect is associated, the authors believe, with the transition of the amorphous alloy under the effect of low-temperature thermal cycling to a new structural state with different short-range parameters. The systematic investigations carried out by the authors of this book show that the main factors,

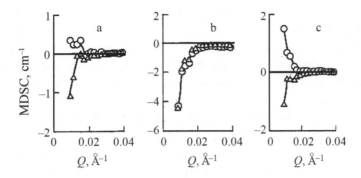

Fig. 3.8. MDSC of the $Ni_{81}P^{19}$ amorphous alloy measured at room temperature prior to (circles) and after (triangles) low-temperature heat treatment. a – specimens 1, b – 3, c – 2.

which determine the extent of changes of the structure and physical properties are the parameters of low-temperature treatment (cooling temperature, duration and the number of cycles) and the composition of the alloys.

The physical model of the low-temperature ΔT-effect is based above all on the examination of the thermal conditions of the process of cooling the specimens of the amorphous alloys. Analysis of the temperature distribution in a three-dimensional plate at any moment of time taking into account the appropriate initial boundary conditions (at the moment $\tau = 0$ all surfaces of the sheet having the temperature T_r are instantaneously cooled to the temperature T_k which is maintained constant throughout the entire cooling process) has the form

$$T(x,y,z,\tau)=(T_rT_k)\sum_{n=1}^{\infty}\sum_{m=1}^{\infty}\sum_{k=1}^{\infty}A_nA_mA_k\left(\cos\mu_n x/R_1\right)\left(\cos\mu_m y/R_2\right)\left(\cos\mu_k z/R_3\right)\times$$
$$\times\exp\left[-\left(\mu_n^2 k_1^2+\mu_m^2 k_2^2+\mu_k^2 k_3^2\right)\alpha\tau/R^2\right],$$

$$(3.8)$$

where $\alpha\tau/R^2$ is the Fourier number, $A_i = (-1)^{i+1}2/\mu_i$, $i = n, m, k$; R is the generalised dimension, $R^{-2}= (R_1)^{-2}+ (R_2)^{-2} + (R_3)^{-2}$; $\mu_i = (i-1)$ $\pi/2$; $ki = R/R_i$, R_1 is the length of the sheet, R_2 is its width, R_3 its thickness. Substituting the typical numerical values into (3.8) it may be seen that the cooling rate of the sheet $V \sim 10^4-10^5$ K·s^{-1} is comparable with the quenching rate (10^6-10^8 K·s^{-1}), obtained in quenching from the melt to produce the amorphous state of the alloys. Such a high value of V is explained by the fact that the thickness of the ribbon is only 20–30 μm. The results of numerical solution of the heat conductivity equation with the identical initial and boundary conditions coincide, as expected, with the exact solution of (3.8).

Equation (3.8) shows that to determine the thermoelastic stresses, taking into account the geometry of the problem $(R_1 \sim R_2 \gg R_3)$, it is sufficient to investigate the one-dimensional problem, namely an infinitely thin sheet. The components of the stress tensor different from zero are equal to [13]

$$\sigma_{11} = \sigma_{22} = 2G\alpha_0 \left[T(z,\tau) - T_r \right](1+\nu)/(1-\nu), \qquad (3.9)$$

where the Z axis is perpendicular to the surface of the sheet, σ_{11} and σ_{22} are the components of the stress tensor acting on the unit areas, normal to each other and parallel to the Z axis; G is the shear modulus, α_0 is the thermal expansion coefficient, ν is the Poisson coefficient. The thermoelastic stresses move with the speed of movement of the isotherms and their magnitude is $\sigma_{11} = \sigma_{22} \approx 10^7$ N · m^{-2}. As shown by the estimates, these stresses are sufficient to break the bonds in the atomic complexes (associates) of the type of borides, phosphides, silicides, etc, which are present in the melt and transfer to the amorphous state during melt quenching. Analysis of the solution of the cooling process of the sheet, described by equation (3.8) shows that at the moment of complete cooling of the sheet the average variation of the volume and the average speed of travel of the surfaces of the sheet differ from zero. This shows that at the moment of complete cooling longitudinal oscillations excited in the sheet and are described by the system of equations [14]

$$\partial D / \partial x - (1-\nu)\partial \Omega / \partial y = \rho \left[(1-\nu^2)/E \right] \partial^2 u / \partial t^2, \qquad (3.10)$$

$$\partial D / \partial y - (1-\nu)\partial \Omega / \partial x = \rho \left[(1-\nu^2)/E \right] \partial^2 \upsilon / \partial t^2,$$

where the volume expansion is $D = \partial u/\partial x + \partial \upsilon/\partial y$; rotation $\Omega = (\partial \upsilon/\partial x - \partial u/\partial y)/2$; u and υ are displacements; ρ is density. Taking into account the initial and boundary conditions and after separating the variables D and Ω it may be seen that two types of longitudinal oscillations form in the sheet: tension–compression oscillations and shear oscillations. The solutions, obtained in the analytical form, can be used to link the frequency of intrinsic longitudinal oscillations of the sheet with its geometrical dimensions and mechanical characteristics of the amorphous alloys:

$$\omega_{1n,m} = \left[E / 2\rho\left(1 - v^2\right) \right]^{1/2} \sum_{n=1}^{\infty} \sum_{m=1}^{\infty} \left[\left(n/R_1\right)^2 + \left(m/R_2\right)^2 \right]^{1/2}; \quad (3.11)$$

$$\omega_{2n,m} = \left[E / 2\rho\left(1 - v^2\right) \right]^{1/2} \sum_{n=1}^{\infty} \sum_{m=1}^{\infty} \left[\left(n/R_1\right)^2 + \left(m/R_2\right)^2 \right]^{1/2}, \quad (3.12)$$

where $\omega_{m,n}$ are the intrinsic frequencies of the tension–compression oscillations; $\omega_{2m,n}$ are the intrinsic frequencies of the shear oscillations.

Thus, the above considerations show that the thermoelastic stresses can cause failure of the atomic complexes, and the oscillations of the sheet are the driving force of the drift of the atoms, for example, by the mechanism proposed by Eiring [15]. The combination of these processes results in the change of the short- and medium-range orders in the amorphous state, causes its homogenising and, in the final analysis, stimulates the transition to the state of the new metastable equilibrium of the amorphous metallic system.

The model of the low-temperature Δ*T*-effect will be used to explain the experimental data. The entire set of the results, obtained by neutron diffraction and Mössbauer spectroscopy indicates the change of the short-range order and homogenising of the amorphous matrix. In particular, the Mössbauer studies showed the change of the local environment of the Fe atoms which after low-temperature heat treatment is enriched with the metalloid atoms. This is an important argument in favour of homogenising of the amorphous structure. The extent and special features of structural changes, like of the physical properties, depend, as expected, on the parameters of low-temperature heat treatment and the composition of the amorphous alloy.

This is confirmed most convincingly by thermal investigations, indicating a decrease of both the low-temperature and high-temperature parts of the activation energy spectrum. The irreversible change of the energy state of the conductivity electrons after the complete cycle of low-temperature heat treatment, observed most accurately by the method of the equatorial Kerr effect, is used as a confirmation of the irreversibility of the processes of structural rearrangements and the transfer of the amorphous alloys to a new structural state under the effect of low-temperature heat treatment.

The decrease of the flow stress (yield stress) and coercive force takes place evidently only in the case of the decrease during low-temperature heat treatment or complete disappearance of the fields of internal stresses in the amorphous matrix.

In high-temperature annealing $[T_a = T_C + (7-10)$ K$]$, where T_a and T_C is the annealing temperature and the Curie temperature of the amorphous alloy, respectively, and the treatment including the low-temperature heat treatment and then annealing at $T = T_a$, the relative changes of the magnetic characteristics of the hysteresis cycle are the additive sum of both processes. This correlates with the result according to which the structural relaxation at $T = T_a$ causes a small decrease of the extent of heat generation in the region of the left branch of the high-temperature part of the activation energy spectrum without influencing at all the high-energy processes, and a special feature of low-temperature heat treatment is the decrease of the amount of heat generated throughout the entire spectrum, including the high-temperature part. Investigations of the main parameters of the hysteresis cycle of alloy f show that the presence in the initial state of a small amount of the crystalline phase has almost no effect on the magnitude of the low-temperature ΔT-effect. This indicates that the thermoelastic and dynamic stresses are evidently not sufficient to cause failure of the nanocrystalline regions in the amorphous matrix and reduce the internal stresses.

The physical model of the low-temperature ΔT-effect proposed in this chapter may be used to explain the resultant experimental data and can be used as a basis for describing the microscopic mechanisms, responsible for the irreversible changes of the structure and physical properties of the amorphous alloys after low-temperature heat treatment.

References

1.   Glezer A.M., Molotilov B.V. Structure and mechanical properties of amorphous alloys. - Moscow: Metallurgiya, 1987.
2.   Amorphous metallic alloys / ed. F.E. Luborsky. - Moscow: Metallurgiya, 1987. - 583 p.
3.   Zhukov A.R., Shtangeev B.L., J. Appl. Phys. - 1993. - V. 73. - P. 5716.
4.   Vinnichenko K.Yu. et al. Proc. Seventh All-Russia Conf. Amorphous precision alloys. - Moscow, 2000. - P. 68.
5.   Zaichenko S.G., et al., Dokl. RAN. - 1999. - V. 367. - P. 478.
6.   Glezer A.M., et al. Izv. RAN. Ser. Fiz. - 2001. - V. 65. - No. 10. - P. 1472.
7.   Glezer A.M., Zaichenko S.G., Izv. RAN. Ser. Fiz. - 2003. - V. 67. - No. 6. - P. 823.
8.   Zaichenko S.G.,et al. Zavod. Lab. - 1989. - No. 5. - P. 76.

9.   Zaichenko S.G., Kachalov V.M., Functional Mater. - 1995. - V. 2. - P. 232.
10. Zaichenko S.G., et al., Abstracts 10th Int. Conf. Rapidly Quenched and Metastable Materials. - Bangalore, 1999. - V. 1. - P. 58.
11. Zaichenko S.G., et al., Proc. 18th Intern. School New magnetic materials for micro-electronics. - Moscow State University, 2000. - P. 70.
12. Calvo-Dahlborg M., et al. Proc. rep. Seventh All-Russia. Conf. Amorphous precision alloys. - Moscow - 2000. - P. 67.
13. Nowacki W. Questions of thermoelasticity. - Moscow: Publishing House of the USSR Academy of Sciences, 1967.
14. Lyav A. Mathematical Theory of Elasticity. - Moscow and Leningrad: Ob'ed. nauchn.-tehn. publ NKTP USSR, 1935.
15. Eiring N., Chem. Phys. - 1936. - V. 4. - P. 283.

The ductile–brittle transition phenomenon

4.1. Main relationships

The authors consider it very important to present a separate chapter dealing with the very important phenomenon associated with the brittle fracture or amorphous alloys and the structural relaxation phenomenon. In this phenomenon, on reaching a specific temperature of preliminary annealing T_{br} within the range of stability of the amorphous state $(T_{br} < T_{cr})$ the amorphous alloys become brittle, completely or partially, at room temperature. This phenomenon of the loss of ductility of the amorphous alloy is not only of purely scientific but also considerable practical interest. It is the measure of relaxation processes in the amorphous alloys and the indicator of their thermal–time stability, and also restricts to a large extent the temperature range of the heat treatment of the industrial alloys which, for example, in the case of magnetically soft materials may include the temperature range close to T_{cr}.

Figure 4.1 shows the typical standard embrittlement curve obtained as a result of mechanical bend tests at room temperature for the Fe–B amorphous alloy annealed at different temperatures [1]. It may be seen that catastrophic embrittlement starts in a very narrow annealing temperature range were the value of the parameter ε_f (ductility in the bend test) changes from unity to almost zero. Consequently, this defines the critical temperature of embrittling annealing T_{br}, and the exact value of this temperature corresponds to the halving of the value of ε_f in relation to the initial state.

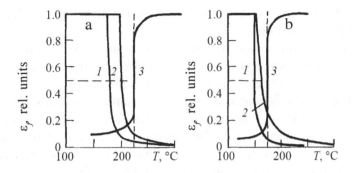

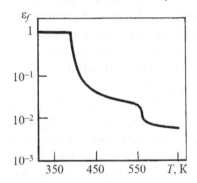

Fig. 4.1. Dependence of the plasticity parameter ε_f on preliminary annealing temperature in vacuum for Fe–B (a) and Fe–B– Ce (b) alloys: curve 1 – annealing time 6 h; curve 2 – annealing time 1 h; curve 3 – dependence $\delta(T)$ giving the value of equicohesive temperature T_e.

Fig. 4.2. Dependence of preliminary annealing (semi-logarithmic scale) for $Fe_{40}Ni_{40}P_{14}B_6$ alloy (annealing time 0.25 h).

More accurate measurements show [2] that the ductility of a ribbon of an amorphous alloy after such a sharp jump remains low but different from zero and there is a second stage of the jump-like decrease of ductility at a higher preliminary annealing temperature.

Figure 4.2 shows the standard curve for the $Fe_{40}Ni_{40}P_{14}B_6$ amorphous alloy with the value ε_f is given on the logarithmic scale. It is therefore possible to record more accurately the two stages of the embrittlement process during annealing. However, it should be remembered that the significance of the first low-temperature stage of embrittlement in which the value of ε_f decreases from 1.0 to 0.1 is considerably greater than the significance of the second high-temperature stage in which ε_f changes within a few hundredths. In subsequent considerations, T_{br} will be regarded as annealing temperature resulting in low-temperature and considerably more significant embrittlement than that shown in Fig. 4.1. In addition to this, it should be remembered that the variation in the range ±0.1 is within the limits of the confidence range in the measurement of bending ductility because the second embrittlement stage is difficult to see in the experiments.

The term temper embrittlement, used most frequently for describing this phenomenon does not describe completely accurately the nature of the process. The authors believe that it is more accurate to refer to this phenomenon as the temper brittleness of the amorphous alloys taking into account the very close analogy between the brittleness formed under the thermal effects in amorphous alloys,and the irreversible temper brittleness in crystalline steels [2]. It is important to stress that this similarity is observed not only in general features but also in very significant details.

Summing up the results of experimental investigations of this phenomenon, it is important to mention the following main relationships of temper brittleness of the amorphous alloys:

1. Every amorphous alloy is characterised by the characteristic embrittlement temperature T_{br} which correlates in a number of cases with the crystallisation temperature in heating the amorphous alloy T_{cr}. Figure 4.3 shows the concentration dependence of the values of T_{br} and T_{cr} in $Fe_{80-x}Ni_xSi_6B_{14}$ alloys [1]. The values of T_{br} were determined by a series of annealing experiments followed by the mechanical bend test at room temperature, and the values of T_{cr} were determined by calorimetric measurements in continuous heating at a rate of 20°/min by recording the temperature corresponding to the crystallisation peak. It may be seen that when varying the nickel content in the amorphous alloys, the behaviour of the values of T_{br} and T_{cr} is the same with the exception of the case when $x = 40$. The experimental results show that the value T_{br} of the amorphous alloys

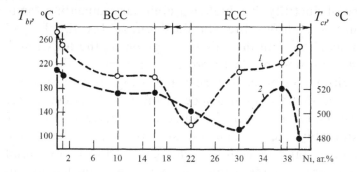

Fig. 4 .3. Dependence of the temper brittleness temperature (curve 1) and the crystallisation temperature (curve 2) of the $Fe_{80-x}Ni_xSi_6B_{14}$ alloys on the nickel content, the arrows indicate the concentration ranges of the existence of the crystalline phases with the BCC and FCC lattices which form during crystallisation of the amorphous state.

based on iron and cobalt decreases in alloying with chromium and molybdenum [3]. The simultaneous presence in the alloy of two types of metalloid atoms, characterised by different characteristics of bonding with the iron atoms, also results in a decrease of T_{br}. This takes place, for example, in the Fe–B–Si, Fe–B–P and Fe–P–C systems [4]. The amorphous alloys containing phosphorus are embrittled at a considerably lower temperature than the alloys containing boron [5] and the simultaneous presence of phosphorus and boron in the alloy results in an even larger decrease of T_{br} [6]. The value of T_{br} slightly increases when boron is replaced by silicon or carbon in the $Fe_{80}B_{20-x}N_x$ system (N = C, Si, P, Ce), and any replacement of boron by phosphorus or cerium reduces the value of T_{br}. In [7] the authors noted the opposite effect of the chemical composition on the value of T_{br} in the Fe–Me–B and Ni–Me–B alloys (Me = Ti, Zr, Hf, Hb, Hb, Na). On the basis of these results it was concluded that the tendency for embrittlement becomes stronger with increasing binding force between the elements including the composition of the amorphous alloys. It is important to know that the temper brittleness is typical not only of the amorphous alloys of the metal–metalloid type but also the metal–metal amorphous alloys [8] (Fig. 4.4).

2. T_{br} is a function of the logarithm of the duration of embrittling annealing and decreases with increase of the duration at the rate which depends on the composition of the amorphous alloy. For example, the value of T_{br} for the $Fe_{82.5}B_{17.5}$ amorphous alloy is 185°C at the annealing time of 6 h and 250°C at the annealing time of 1 h [1]. Undoubtedly, this indicates the thermally activated nature of embrittlement. Figure 4.5 shows the dependence of the annealing time to embrittlement (on the logarithmic scale) on the inverse temperature for the amorphous alloys based on Fe–Si–B and Fe–Si–Cr–B [9].

It can be seen that the Arrhenius law is satisfied in this case by the process temper brittleness and in principle it is possible to determine the effective activation energy of the embrittlement process. The two values of the activation energy obtained, for example, for the $Fe_{40}Ni_{40}P_{14}B_6$ amorphous alloy are 0.94 eV (for the low-temperature main embrittlement stage) and 2.47 eV (for the high-temperature embrittlement stage) [10]. It is very important that the activation energy of 0.94 eV is too small to cause diffusion of atoms over large distances. Identical results for the amorphous alloys of the metal–metal type (Ni–Nb–Al) were obtained by the authors of [11]. The

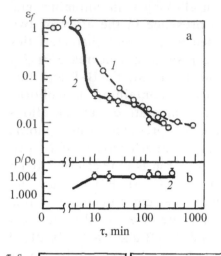

Fig. 4.4. Variation of the ductility parameter ε_f (a) and the relative density ρ/ρ_0 (b) in dependence on the annealing time at 350°C in the $Ni_{61}Nb_{39}$ (1) and $Ti_{50}Be_{40}Zr_{10}$ (2) alloys.

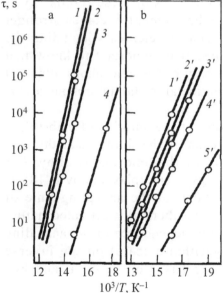

Fig. 4.5. Dependence of the annealing time to embrittlement (logarithmic scale) on the inverse annealing temperature for Fe–Si–B (a) and Fe–Cr–Si–B (b) alloys: curve 1 – $Fe_{81}Si_8B_{11}$ alloy ($H = 310$ kJ/mol); curve 2 – $Fe_{79.5}Si_{8.5}B_{12}$ alloy ($H = 309$ kJ/mol); curve 3 – $Fe_{78}Si_9B_{13}$ alloy ($H = 269$ kJ/mol); curve 4 – $Fe_{75}Si_{10}B_{15}$ alloy ($H = 227$ kJ/mol); curve 1' – $Fe_{71}Cr_{10}Si_8B_{11}$ ($H = 1170$ kJ/mol); curve 2' – $Fe_{69}Cr_{10}Si_9B_{12}$ alloy ($H = 174$ kJ/mol); curve 3' – $Fe_{72}Cr_{10}Si_8B_{10}$ alloy ($H = 166$ kH/mol); curve 4' – $Fe_{68}Cr_{10}Si_9B_{13}$ alloy ($H = 148$ kJ/mol); curve 5' – $Fe_{65}Cr_{10}Si_{10}B_{15}$ alloy ($H = 120$ kJ/mol); H is the effective activation energy of temper brittleness proces.

activation energy spectrum of temper brittleness determined in [11] does not correspond to the occurrence of the diffusion processes and, consequently, it may be assumed that these processes are controlled by the local atomic rearrangement in the structural relaxation stage.

3. Brittleness is typical of certain amorphous alloys in the initial condition; at the same time, there are alloys with the amorphous structure which shows no signs of temper brittleness up to crystallisation (in most cases, these are amorphous alloys of the metal–metal type). In reality, this means that the structural

state, typical of the amorphous alloy after embrittling annealing, is capable of forming directly after melt quenching. However, this structural state cannot form under thermal effects which do not change amorphous state. It is shown that the parameters of formation in melt quenching (quenching rate, the superheating temperature of the melt) not only have a strong effect on the value of T_{br} [12] but in a number of cases determine the very possibility of producing amorphous alloys in the initial ductile state [13]. The general trend is such that a decrease of the quenching rate results in a decrease of the values of T_{br} of amorphous alloys [14].

4. The value of T_{br} rapidly decreases when the amorphous alloys are alloyed with surface-active elements. The experimental results show [15] that elements such as Te, Se or Sb have a strong embrittling effect on the amorphous alloys based on iron and nickel and decrease the value of T_{br} even if they are added in small amounts (Fig. 4.6). In addition to this, Fig. 4.1 shows the results of determination of T_{br} using the dependence $\varepsilon_f(T_{ann})$ for the Fe–B alloy [1]. Adding only 0.01 at.% Ce results in a large decrease of T_{br}, by approximately 50°C.

4.2. Plastic flow and temper brittleness

It is of considerable interest to compare the values of T_{br} with the characteristic temperature of transition from the inhomogeneous mechanism of plastic flow to the homogeneous mechanism. This characteristic (the so-called equicohesion temperature) is typical of every amorphous alloy and slightly changes in dependence on the method and the rate of deformation. It is well-known that

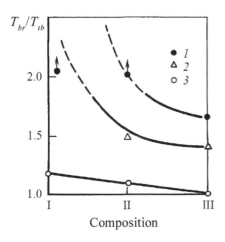

Fig. 4.6. The value T_{br}/T_{tb} in dependence on the initial composition and the nature of the surface-active addition added to the basic alloy in the amount of 0.1 at.%: T_{br} is the temperature of temper brittleness of the basic alloy; T_{tb} is the temperature of temper brittleness of the same alloy with the addition [320]; curve 1 – Te, curve 2 – Se, curve 3 – Sb, I – $Fe_{81.5}B_{14.5}Si_4$; II – $Fe_{40}Ni_{40}B_{20}$; III – $Ni_{81.5}B_{14.5}Si_4$.

the transition through the point T_e during deformation results not only in the transition to the viscous homogeneous flow but also subsequent embrittlement in the room temperature range. To compare the values of T_{br} and T_e, tests were carried out by uniaxial tensile loading of the Fe–B amorphous alloys at different temperatures. The resultant dependence of the relative elongation to fracture on the test temperature is also shown in Fig. 4.1. It may be seen that the values of T_e are higher than those of T_{br}, and the decrease of one of these temperatures after adding the surface-active microaddition causes the same change of the other characteristic temperature. The experimental results can be used to conclude that the occurrence of the homogeneous plastic flow at elevated temperatures should always result in subsequent embrittlement because heating is accompanied by the transition through the temper brittleness temperature.

The results presented above will now be compared with the data for the effect of the surface-active elements on the dependence of the parameter δ characterising the susceptibility of the amorphous matrix to plastic flow (Fig. 4.7). In may be seen that, firstly, δ remains almost constant in the temperature range corresponding to the development of temper brittleness. Secondly, the addition of antimony or cerium, which greatly reduces the value of T_{br}, results in a small decrease of δ. Evidently, the decrease is not caused by the surface-active nature of this element because the addition of small amounts of the element such as niobium results in an even larger decrease of δ (Fig. 4.7). A large decrease of δ is recorded for all Fe–B alloys in the annealing temperature range 300–350°C, with these

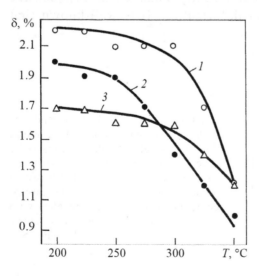

Fig. 4.7. Effect of microalloying with surface-active elements on the dependence of ductility at room temperature on the preliminary annealing temperature for 6 hours for the $Fe_{83}B_{17}$ alloy: curve 1 – no addition; curve 2 – addition of antimony; 3 – addition of niobium.

values being considerably higher than T_{br} (by approximately 100°C) and corresponding to the high rate of formation of the clusters – pre-precipitates of the crystalline phase.

Another important question is the presence or absence of changes in the nature of the local plastic flow at room temperature with increasing preliminary annealing temperature. The experimental results show [1] that after heat treatment in different conditions the density of the shear bands and the average length of these bands in the uniaxial tensile bend test do not change significantly. In particular, there were no changes in the parameter of distribution and geometry of the bands after heat treatment in the range of T_{br}. The average local thickness of the shear bands also does not change in the temperature range corresponding to the occurrence of temper brittleness.

The main results of the analysis of the acoustic emission spectra of the amorphous alloys based on iron and cobalt show [15] that in the uniaxial tensile test both the freshly quenched and thermally embrittled ribbons retained the capacity for plastic deformation. This is also confirmed by the presence of the shear bands with adjacent steps (according to the scanning electron microscopy results). On the time dependence of acoustic emission the pulses appeared from the moment of the start of plastic flow, the average time between the pulses was 0.25 s for the freshly quenched material and 1.1 s for the embrittled material. The pulses of the same amplitude also differed in the duration: in the case of the freshly quenched material the duration of the pulses changed in the range $(300–500) \cdot 10^{-6}$ s, in the embrittled alloy in the range $(400–600) \cdot 10^{-6}$ s (Fig. 4.8). The statistical analysis of the time distribution of the pulses shows that for the freshly quenched specimens this distribution corresponded to the Poisson distribution [16]. Comparison of the number of the shear bands with the number of pulses shows that they are in agreement. Thus, each pulse corresponded practically to the formation of a single shear band and the pulse duration corresponded to the time of its formation. If the pulses accompanying shear deformation have many similar features, the main difference for the embrittled material was the presence of two or three powerful pulses following each other immediately prior to failure of the ribbons. The energy of these pulses was two orders of magnitude greater than the energy of the pulses corresponding to the shear strains, and characterised the crack formation processes prior to the final failure of the ribbon. In contrast to the shear pulses, characterised by a truncated envelope (due to internal filling with discrete frequencies), the filling of the latter was continuous (no discrete frequencies). In the case of the embrittled specimens it is possible to discuss only the microplasticity of the individual sections because the total number of the pulses in the specimens was an of magnitude smaller than in the freshly quenched specimens [16].

The interpretation of the acoustic emission spectra together with the scanning electron microscopic data for the width of the steps

In the graph: mB = mV; mkc = μs

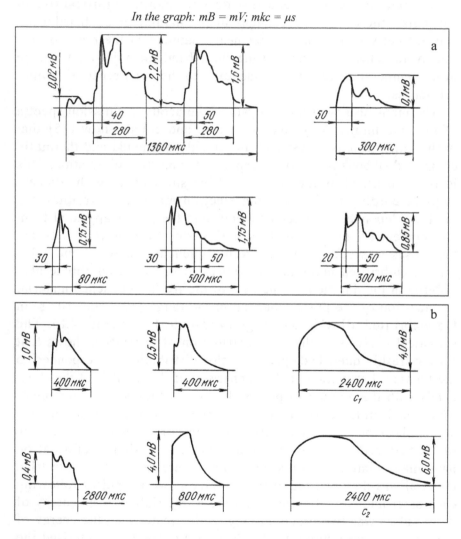

Fig. 4.8. Acoustic emission pulses, accompanying deformation of the iron-based amorphous alloy and its fracture in the quenched (a) and embrittled (b) states; c_1 and c_2 are the pulses characterising crack formation prior to failure.

was carried out to determine the rate of formation of the shear bands equal to $7.8 \cdot 10^{-5}$ m/s which is comparable with the lower estimate of this parameter obtained by high-speed filming ($4 \cdot 10^{-5}$ m/s [17]).

The structural relaxation stage, preceding temper brittleness, also did not show any changes in the fractographic special features. A characteristic feature was the combination of the areas of almost smooth cleavage with a groove pattern formed by a system of veins. The only large difference in the fractographic features after annealing at temperatures close to T_{br} was the appearance of micropores in the alloys in which they were not detected in the initial state, and the increase of the density of the micropores in the amorphous alloys Fig. 4.9). After annealing leading to embrittlement the mechanism of nucleation and growth of cracks changed qualitatively.

The results of electron microscopic experiments show [1] that in this case the cracks nucleated at the micropores whose number and size greatly increased as a result of the effects (Fig. 4.10). The region of local plastic flow was clearly detected ahead of each crack formed at the pores (Fig. 4.10). This again confirms the previously made conclusions according to which the susceptibility to plastic flow in the amorphous matrix is still very high and the cleavage cracks are in fact quasi-brittle cracks, regardless of the macroscopic brittle nature of fracture. The micropores in the amorphous matrix are not only sources of microcracks but also facilitate their subsequent growth determining in a number of cases the trajectory of propagation of the main cracks (Fig. 4.10).

Discussing the changes in the structure of the amorphous alloys in the stage in which embrittlement takes place, it should be concluded

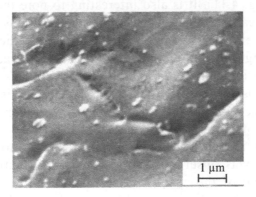

Fig. 4.9. Fracture surface of the Fe–B amorphous alloy annealed at temperatures higher than T_{br} (scanning electron microscopy in reflected electrons).

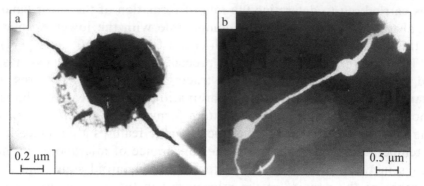

Fig. 4.10. Nucleation (a) and propagation (b) of cracks at submicropores in Fe–B alloy; dark field (a) and bright field (b) images produced by TEM.

that no changes were detected in the structure using direct diffraction experimental methods [18]. The changes in the structure are not manifested in the radial distribution function or correlation functions or electron microscopic images in the mode of formation of the amplitude or phase contrast. This method of low-angle scattering of the X-rays results in a large shift in the nature of distribution and the average size of the regions of the free volume [19]. As shown by the dilatometric experiments, the values of T_{br} coincide with the temperature range of the most active occurrence of the processes of densening of the amorphous matrix as a result of annihilation of the excess free volume. Analysis of the data, obtained for the specimens of the amorphous alloys based on Fe–B, annealed for six hours at 200 and 300°C, indicates that there is the regular shift of the size distribution of the regions of the free volume in the direction of higher values (Fig. 4.11). It is also interesting to note the successive increase during heat treatment of the individual fractions of the areas of the free volume: the first stage is characterised by the increase of the statistical weight of the intermediate (40±15) nm fraction so that the curves of the indicatrix invariants show new peaks; the second stage is characterised by either the increase of the statistical weight of the largest fraction or by the formation of another distribution peak in the area of large defect sizes.

It was noted in [20] that the brittleness in annealing the $Fe_{27}Ni_{53}P_{14}B_6$ alloy is observed in the temperature range in which the largest change of the specific electrical resistivity and Curie temperature takes place. Consequently, it may be concluded that the brittleness is associated with the formation of a more stable short-range order structure in the amorphous matrix.

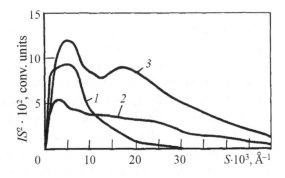

Fig. 4.11. Indicatrix invariants of low-angle scattering in dependence on the structural state of the $Fe_{83}B_{17}$ alloy: curve 1 – quenched state; curve 2 – annealing at 200°C, 6 h; curve 3 – annealing at 300°C, 6 h.

4.3. Analysis of structural models of temper brittleness

To explain the nature of temper brittleness of the amorphous alloys the experimental results were used to propose two groups of models:

1. 'Segregation' model [21], explaining the brittleness by the formation of segregations of the atoms–metalloids in certain areas of the amorphous matrix.

2. 'Crystalline' model [22] which links the brittleness with the formation in the amorphous matrix of a significant short-range order or crystalline phases of a specific type.

Both models are based on a certain set of indirect experimental data but cannot explain a number of experimental facts confirming the alternative model or not fitting in any of the two models. For example, the 'segregation' model is not capable of explaining the presence of temper brittleness in the amorphous alloys of the metal–metal type. At the same time, the 'crystallisation' model contradicts the data for the rapid decrease of the value of T_{br} observed when adding small amounts of surface-active elements. In most cases, the conclusions regarding the validity of a specific model are 'structureless'. They were proposed not on the basis of the detailed structural studies of the relationships governing the plastic flow and failure in transition through T_{br}, and were based mostly on the results of examination of a number of external influences on the given characteristic followed by not always correct conclusions in which the desirable result was often treated as the actual result. In a number of experiments (for example, [23]) in discussing the results relating to the temper brittleness phenomenon, the authors presented more accurate but at the same time less physically justified assumption according to which the temper brittleness is associated with the processes of structural relaxation in the amorphous structure. This is supported by the agreement of the temperature ranges of structural relaxation and occurrence of temper brittleness observed in the studies (on alloys of the Fe–B, Fe–P, Fe–Ni–P, Fe–Ni–P–B systems) and also by similar values of the activation energy of structural relaxation and embrittlement (for the $Fe_{40}Ni_{40}P_{14}B_6$ alloy). However, in these cases none of the studies proposed a specific mechanism by which the structural relaxation could result in extensive embrittlement from the position of the physics of plastic deformation and fracture.

Attention will be given in greater detail to the 'segregation' and 'crystallisation' models of temper brittleness in the light of the results obtained for relaxation hardening together with standard embrittlement curves. Figure 4.12 shows the dependence of the parameters ε and δ and also of the yield strength σ_T for the $Fe_{83.5}B_{17.5}$ alloy on the preliminary annealing temperature ($\tau_{ann} =$ 1 h). Detailed combined analysis of these relationships leads to the following conclusions:

1. The segregation processes determining the effect of low-temperature hardening (LTHE) (the first maximum of the $\sigma_T(T_{ann})$ curve) cannot be the only reason for the formation of temper brittleness because T_{LTHE}, corresponding to the first maximum, is considerably lower than T_{br}. A similar relationship between T_{LTHE} and T_{br} is characteristic of the majority of the investigated amorphous alloys in which the effect of low-temperature hardening was recorded.

2. The high degree of the topological short-range order – the stage of pre-precipitation of crystalline phases – also does not cause temper brittleness because T_{LTHE}, corresponding to the second maximum on the $\sigma_T(T_{ann})$ is considerably lower than T_{br}. In particular, in the Fe–B alloy the difference between T_{LTHE} and T_{br} is approximately 100°C.

In the investigations, defending the 'crystalline' model of temper brittleness, it is assumed that the value of T_{br} in the amorphous alloys where the transition to the crystalline state takes place with the formation of the BCC phase, is considerably lower than in the alloys in which the FCC phase forms as a result of the formation of internal stresses formed during the nucleation of the BCC phase in the amorphous matrix [24]. To verify this assumption, attention will be given again to Fig. 4.3 which shows the concentration dependences of T_{br} and T_{cr} for the $Fe_{80-x}Ni_xSi_6B_{14}$ amorphous alloys.

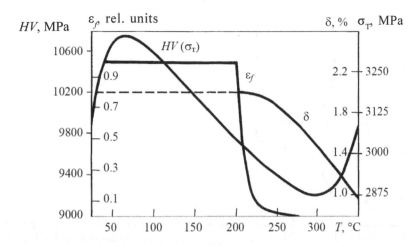

Fig. 4.12. Comparison of the dependences of the yield limit on ductility ε_f and the plastic flow susceptibility δ on the temperature of preliminary annealing of $Fe_{83}B_{17}$ alloy for 1 hour.

The identification of the first crystalline phases by the method of transmission electron microscopy, presented at the same time, shows that the nickel concentration of approximately 20 at.% results in the change of the crystal lattice of the precipitated metallic phase from BCC to FCC. This has no effect on the concentration dependence of T_{br}. Obviously, this indicates that the model proposed in [24] is incorrect and also confirms to some extent the important conclusion according to which the temper brittleness forms particularly in the amorphous state.

Thus, neither the 'segregation' nor 'crystalline' model can be used to explain the phenomenon of temper brittleness. At the same time, the relaxation nature of this phenomenon is clearly evident. In order to confirm again this suggestion, the dependence $\sigma_T(T_{ann})$ will again be compared with the temperature dependence of the thermal expansion parameter $\Delta l / T$, where Δl is the variation of the linear size of the specimens and T is absolute temperature (Fig. 14, see also Fig. 4.4). The graph shows the dependences obtained for the Fe–Ni–P amorphous alloy [1], but similar dependences can also be obtained for other alloys. It may be seen that the value of T_{br} coincides with the temperature range of the most active occurrence of the relaxation processes which lead to considerable densening of the amorphous structure.

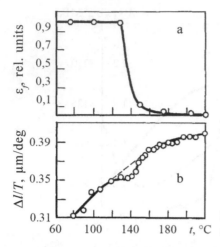

Fig. 4.13. Comparison of the dependences $\varepsilon_f(T_{ann})$ (a) and $\Delta l/T$ (b) for the Fe–Ni–P alloy.

4.4. Ductile–brittle transition from the viewpoint of mechanics of plastic deformation and fracture

Taking into account the general considerations regarding the nature of the brittle state, it is possible to specify two possible processes which may result in the loss of the macroscopic plasticity of the amorphous alloy as a result of thermal effects: 1) decrease of the susceptibility to plastic flow, and 2) easier process of formation and subsequent growth of the cracks. The susceptibility to plastic shape changes in the amorphous alloys remains almost completely constant after annealing at temperatures higher than T_{br}. In particular, this results from the fact that the value of parameter δ after these heat treatments is close to the appropriate value for the alloy in the initial condition and does not show any significant changes in the range of T_{br} (Fig. 4.7).

Consequently, the phenomenon of temper brittleness of the amorphous alloys is caused not as much by the suppression of the plastic flow processes as much by easier fracture processes. The reasons for this phenomenon will be examined in greater detail.

To explain the nature of the processes forming the basis of the fracture mechanics of amorphous alloys and playing a significant role in suppressing temper brittleness, it is important to take into account studies by Kimura and Masumoto [25, 26]. They used the criterion of local failure in tensile loading which is capable of describing the mechanisms of the ductile–brittle transition in the amorphous alloys. In a general case:

$$\sigma_f = \sigma_{yy\,max} = K_e K_p\,\sigma, \qquad (4.1)$$

where σ are the stresses applied to the specimen containing a crack; $\sigma_{yy\,max}$ is the maximum longitudinal stress; σ_f is the microfracture stress; K_e and K_p are respectively the stress intensity factor and the plastic strain intensity factor.

Figure 4.14 shows the values of the macroscopic yield limit σ_T, fracture stress σ_{fr}, and the stresses corresponding to the appearance of the first signs of sliding σ_1 for the $Pd_{78}Cu_6Si_6$ amorphous alloy in the plane strain conditions of a notched specimen in dependence on the annealing time of the alloy at 350°C. There is a sharp ductile–brittle transition reflecting the behaviour of the measured characteristics [27].

Figure 4.15 shows the calculated values of the microfracture stress σ_p and the maximum value of the yield limit under the base of the notch $K_{max}\sigma_T$ and the yield stress under uniaxial tensile loading at the base of the notch σ_T in dependence on annealing time at the same temperature.

Comparison of the graphs in Figs. 4.14 and 4.15 shows unambiguously that the loss of plasticity in annealing of the Pd78Cu6Si16 alloy can be explained by a decrease of the microfracture stress and not by making the sliding process more difficult. Thus, the ductile– brittle transition in annealing can be explained by a decrease of the microfracture stress below the maximum value of the yield limit in the longitudinal direction. Consequently, using the fracture criterion, proposed by Kimura and Masumoto, it can also be concluded that the temper brittleness is associated with making the fracture process easier and not with complication of the plastic flow process. In this connection, the characteristic of the mechanical behaviour of the system such as the microfracture stress becomes the most important physical parameter characterising the brittle fracture susceptibility of the amorphous alloys. It has been shown that the microfracture stress of the $Pd_{78}Cu_6Si_{16}$ amorphous alloy is approximately 4.2 GPa which corresponds to the value $E/20$, where E is the Young modulus equal to 89.7 GPa, which is close to the theoretical strength $E/10$. At the same time, the microfracture stress of the $Fe_{40}Ni_{40}P_{14}B_6$ amorphous alloy equals 2.88 GPa [27]. In principle, it is concluded that the ductility of the amorphous alloys is determined directly by the higher value of the microfracture stress in comparison with the amorphous inorganic materials.

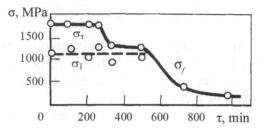

Fig. 4.14. The macroscopic yield limit σ_T, fracture stress σ_f and the stress corresponding to the appearance of the first signs of sliding σ_1 for the $Pd_{78}Cu_6Si_{16}$ alloy in the plane strain conditions in dependence on annealing time at 350°C.

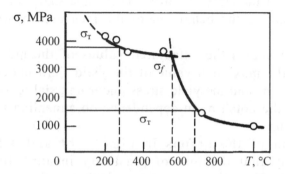

Fig. 4.15. Calculated values of the microfracture stress σ_f and the yield stress σ_T in dependence on the preliminary annealing time; the alloy and the annealing conditions are given in Fig. 4.14.

It is also important to note that the microfracture stress is a structure-sensitive characteristic of the material and tends to decrease rapidly and the thermal effect leading to temper brittleness. What are the structural reasons for the rapid decrease of the microfracture stress in annealing of the amorphous alloys? It will be attempted to answer this question in the following section.

4.5. Ductile – brittle transition and the free volume

Taking into account the previously discussed experimental data and also the discussed thermally activated nature of temper brittleness it may be concluded unambiguously that this phenomenon is of the relaxation nature. The structural relaxation of the amorphous alloys is itself a very complicated phenomenon including several processes connected by complicated relationships and taking place with greatly differing intensity in different temperature ranges. It is therefore necessary to clarify what are the processes of structural relaxation leading to the change of the fracture mechanism and, correspondingly, to a large decrease of the microfracture stress and specific thermal effects.

A number of experimental factors will be taken into account:

1. T_{br} coincides with the temperature range of the most active occurrence of the processes of 'densening' of the amorphous matrix as a result of annihilation of the excess free volume (Fig. 4.13);

2. Increase during structural relaxation of the number and size of the micropores recorded by macroscopic examination methods (Fig. 4.10);

3. Appearance in the stage of temper brittleness of extensive crack formation at the micropores (Fig. 4.9);

4. A large decrease of the value of T_{br} when atoms of tin, antimony or cerium are added to the amorphous alloys leading to qualitative changes of the microfracture mechanisms as a result of the effect on the effective surface energy of the nucleated and growing cracks;

5. The amorphous alloys, produced by spraying and characterised by higher porosity, show considerably lower values of T_{br} in comparison with the alloys produced by melt quenching [28].

6. The parameters of production of the alloy in melt quenching have a strong effect both on the ductility in the initial state and on the value of T_{br}. Figure 4.16 shows the standard curves for $Fe_{40}Ni_{40}B_{20}$ alloy produced in the initial condition in the form of ribbons of different thickness. It may be seen that the thicker initial ribbon, which corresponds to a lower melt quenching rate, has a lower value of T_{br}. With increase of the quenching rate (decrease of the ribbon thickness) the value T_{br} becomes higher. The change of T_{br} is very large and equals approximately 100°C when the thickness of the ribbon is varied from 30 to 50 μm. The curves of variation of the density for the ribbons of different initial thickness (Fig. 4.17) were obtained in Ref. 4. Two circumstances should be noted: 1) as the quenching rate increases the temperature of the most active 'densening' of the amorphous alloy also increases, and 2) the temperature range of rapid 'densening' always coincides with the value of T_{br}.

All these results undoubtedly indicate that the controlling role in the temper brittleness phenomenon is played by the excess free volume and the mechanism of its evolution under the thermal effect on the amorphous structure.

The role of the free volume in the process of propagation of the crack in the amorphous matrix will be investigated. In the absence of

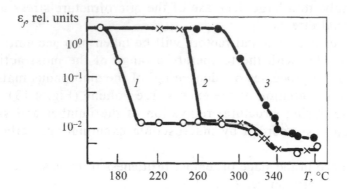

Fig. 4.16. Dependence of the plasticity parameter ε_f on the preliminary annealing temperature for 43 h for $Fe_{40}Ni_{40}B_{20}$ alloy. Ribbons with a thickness of 50 μm (curve 1), 40 μm (curve 2) and 30 μm (curve 3).

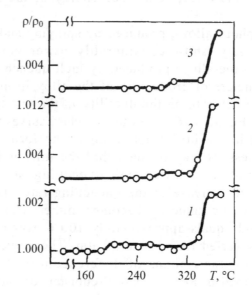

Fig. 4.17. Variation of relative density $\Delta\rho/\rho$ during preliminary annealing for 43 hours at different temperatures of $Fe_{40}Ni_{40}B_{20}$ alloy. The notations are the same as in Fig. 4.16.

relaxation by plastic shear, the stress concentration in the vicinity of the crack tip σ causes the dilation $\Delta V/V$, expressed by the equation [29]:

$$\sigma = E\Delta V / V = \left(E\gamma_s / l \right)^{1/2},   \tag{4.2}$$

where γ_s is the surface energy (the energy of formation of the surface per unit area); E is the Young modulus; l is the crack length.

Spaepen and Turnbull assume [29] that in the amorphous alloy with no crystal lattice similar dilation supports the formation of the excess free volume at the crack tip. This results in a local decrease of the ductility of the matrix. In the free volume model deformation takes place by atomic displacement by diffusion of the free volume. In order for plastic deformation to take place under given conditions (applied stress, temperature, etc), the specimen must contain a sufficient amount of the free volume capable of migration. Thus, within this model, the stress concentration at the crack tip is regarded as a unique source of the free volume playing the role of the dislocations in the crystal. Since the free volume is difficult to measure quantitatively, it is usually estimated by measuring the ductility of the system. The ductility relates the flow speed $\dot{\varepsilon}$ with the local stress σ using the Newton equation:

$$\dot{\varepsilon} = \sigma / 3\eta. \tag{4.3}$$

The mobility of the atoms in the crystalline solids is proportional to the probability of existence of the vacancies in the nearest cell multiplied by the probability of overcoming the energy barrier for the transition of the atom to the area of the vacancy. Therefore, the diffusion coefficient is equal to:

$$D = \nu b^2 \exp\left(-\Delta G_v / kT\right)\exp\left(-\Delta G_m / kT\right), \tag{4.4}$$

where the first exponent determines the concentration of vacancies, and the second exponent the probability of the diffusion transition. Here ΔG_v and ΔG_m is the free energy of respectively the formation and migration of the vacancies; ν is the Debye frequency; b is the atomic spacing. The ductility can be determined using the Stokes – Einstein equation as follows:

$$\eta = kT / Db \approx \left(kT / \nu b^3\right)\exp\left(\Delta G_v / kT\right)\exp\left(\Delta G_m.kT\right). \tag{4.5}$$

In the Spaepen model of deformation by the free volume the ductility of the amorphous alloy below the glass transition point T_c is defined by the equation [30]

$$\eta = kT / \nu b^3 \exp\left(-\delta V * / V_f\right)\exp\left(\Delta G_m / kT\right), \tag{4.6}$$

in which the probability of finding the vacancy together with the atom is replaced by the probability of finding the cavity or free volume with the size V^*. This probability is expressed as $\exp(-\delta V^*/V_f)$ [29], where δ is the geometrical factor of the order of unity. The expressions for the crystals and the amorphous solid are in principle identical with the exception of the fact that in the case of the amorphous state the size distribution of the regions of the free volume replaces the expression for the vacancies in the crystal lattice.

Wu and Spaepen assumed [30] that the ductile–brittle transition temperature (not necessarily temper brittleness) in the given amorphous alloy is associated with some critical value of V_f and, consequently, the critical value of η. In the absence of vacancies, the decrease of the free volume at the equilibrium can be regarded as a consequence of the thermal expansion: $\Delta V_f = \alpha(T-T_0)$. In the process of fast quenching part of this value is captured at temperatures below T_c by the amorphous matrix as in the case of freezing of the vacancies in the crystal. However, part of the free volume can undoubtedly disappear as a result of annealing the amorphous state. Consequently, the free volume concentration V_f at the test temperature is determined by the following relationship

$$V_f = \Delta V_{fs} + \Delta V_{fq} + \Delta V_{fa}\alpha\left(T - T_0\right), \qquad (4.7)$$

here ΔV_{fs} is the variation of the free volume, caused by the stress; ΔV_{fq} is the fraction of the free volume frozen in quenching, ΔV_{fa} is the variation of the free volume in annealing.

Thus, both annealing and decrease of the test temperature decrease V_f, increase η, by increasing the critical stress of the start of plastic flow. These assumptions were also used to carry out detailed calculation of the effect of V_f on the nature of the flow and the relaxation at the stress raisers [31]. The complicated stress relaxation at the crack tip at a restricted concentration of the free volume is in principle capable of decreasing the critical fracture stress or, using the terminology proposed by Kimura and Masumoto, the decrease of the free volume concentration may result in a decrease of the microfracture stress.

In [32] this approach was supplemented by an assumption of the development in the amorphous matrix of the zones enriched with a metalloid. According to the author, the development of the zones of this type is capable of producing in the amorphous matrix areas in which the size of the free volume is considerably smaller than the average size for the entire alloy, and it is they that determine the brittleness in annealing.

The previously discussed concept of temper brittleness, based on the insufficient concentration of the free volume in the amorphous matrix, together with the experimental confirmation (for example, agreement of T_{br} with the temperature range of the largest densening) has three main shortcomings:

1. In accordance with the model proposed by Wu and Spaepen, the stage of temper brittleness should result not only in easier fracture but also more difficult plastic flow. However, the experiments show that this is not the case: the susceptibility to plastic flow in the amorphous alloys, annealed both above T_{br} and below T_{br} is approximately the same.

2. The model proposed by Wu and Spaepen treats the propagation of the already formed crack whose opening is made more easier as a result of the more difficult stress relaxation at the tip of the crack. At the same time, temper brittleness results not only in easier crack opening but also in easier crack initiation and this is not discussed in this model.

3. Using the model, it is difficult to understand why alloys of different composition have different values of T_{br}, and some of them do not show temper brittleness at all.

It has been attempted to simulate the process of stress relaxation around an ellipsoidal pore under uniaxial tensile deformation with a constant rate [31]. It was assumed that the material is a non-linear ductile–elastic continuum with the ductility which depends on local dilation, i.e., on the presence of the free volume. The calculation results show that if the local free volume formed as a result of dilation is not capable of instantaneous redistribution in accordance with the new equilibrium distribution, the tensile stress would results in a disruption of integrity. By introducing a time constant for the process of redistribution, the authors of [31] predicted the ductile–brittle transition in cracking on a micropore.

The nucleation of a crack in the area with a reduced density (in the limit at the micropore) can be easily predicted on the basis of the model proposed in [35]. It is shown in [35] that the area with the reduced value of the elasticity modulus is capable of producing a brittle crack in the interaction of the shear band with the boundary separating the regions with different moduli. The experimental results obtained in [35] show that if the difference of the moduli in the neighbouring areas becomes fourfold, the nucleation of the brittle crack becomes spontaneous under the conditions of external loading. Since the region of the free volume can during its evolution transfer into a submicropore in the stage of high-rate annihilation of the free volume, there is a critical moment when the discontinuity is capable of generating a crack. In this case, the situation can be aggravated even more by the fact that an impurity, reducing the surface energy, may segregate on the interfacial surface of the areas with different density (and, consequently, different elasticity moduli).

The actual possibility of formation of cracks at a micro-discontinuity was shown by the authors of the book in Fig. 4.9. Two conditions for easier crack formation in the temper brittleness stage, combining in fact the relaxation and segregation models of this phenomenon, will be formulated. For example, the following reasons may lead to a rapid decrease of the microfracture stress in the stage of temper brittleness without any large decrease of the susceptibility for plastic flow:

1. The presence of a large number of submicropores with the size larger than critical (depending on the composition of the alloy), ensuring spontaneous crack nucleation at these submicropores.

2. Formation of segregations at the submicropores leading to a decrease of the critical parameters of the spontaneous crack opening displacement at the submicropores. In addition, the decrease of the cohesion ahead of the front of the growing crack (as a result of the inflow of the atoms of the surface-active element from the matrix) leading to easier crack propagation.

The processes of structural relaxation during low-temperature annealing causing temper brittleness and leading to fulfilment of all these conditions will be discussed briefly. The size distribution of the areas of the free volume, typical of each amorphous alloy in the initial condition, depends strongly on the composition of the alloy and the production parameters. At the same time, during structural relaxation in heating the amorphous alloy this distribution changes regularly: it is displaced to larger values as a result of annihilation of the lower (most mobile) free volume regions. In addition, some of the defects transfer to the larger micropores and microdiscontinuities of metallurgical origin. The temperature range of the highest intensity of structural relaxation depends on the parameter of the size distribution of the free volume: the larger the displacement of the size distribution towards higher values the lower are the annealing temperatures at which high intensity structural relaxation processes take place. Thus, each amorphous alloy with a specific composition and the size distribution of the free volume regions will be characterised by a specific temperature (and duration) in annealing leading to the formation of the number of the micropores of the 'supercritical' size sufficient for the appearance of temper brittleness. When the critical size is determined mainly by the chemical nature of the components of the amorphous alloy, and the size distribution by the production parameters, different values of T_{br} may be obtained: from room temperature (the alloy fails by brittle fracture in the initial state) to the crystallisation temperature (no temper brittleness), and this was also observed in practice in particular, a decrease of the quenching rate greatly displaces the size distribution of the defects to higher values and this also results in the experimentally observed decrease of T_{br}.

This model can be used to understand easily the dependence of the value of T_{br} on the method of testing plasticity by bending: when the free surface is located in the zone of tensile stresses (on the external side) the value of T_{br} is lower than in the case when the zone of tensile stresses contains the contact surface of the ribbon of the amorphous alloy. The specific features of the effect of shaping forces in quenching on a cooling disc is such that the contact surface is characterised by slightly higher (in comparison with the average value in the volume) concentration of smaller defects. At the same time, a higher concentration of larger submicropores forms on the external surface and this increases the probability of brittle fracture and under the effect of bending forces when the free surface is located in the zone of the effect of tensile stresses.

It is important to mention another important aspects of the fracture of the amorphous alloys in the temper brittleness conditions – the mutual effect of the plastic deformation and fracture processes. The plastic flow susceptibility of the amorphous alloys in the temper brittleness state is as high as in the initial state. In other words, the plastic flow does not suppress the fracture processes. This fact, paradoxic for the crystalline material, is fully regular for the amorphous state. The optimum propagation path for the crack is in the area of the amorphous matrix in which plastic shear took place because the shear band is characterised by a smaller local thickness and lower density in comparison with the surrounding non-deformed amorphous matrix. Consequently, the plastic flow not only does not inhibits the quasi-brittle fracture but also stimulates it to a certain extent. The surface-active elements in turn influence the creation of conditions for the development of temper brittleness because of two reasons: firstly, by displacing in melt quenching the size distribution of defects in the direction of higher values and, secondly, they are adsorbed in the regions of the free volume and, consequently, reduce the effective critical size for the propagation of the cracks. In addition, decreasing the cohesion, the surface-active elements support the propagation of brittle fracture. All these factors lead to the previously mentioned rapid decrease of T_{br} when adding microadditions of antimony, tin and cerium.

From the viewpoint of the relaxation model of the free volume, discussed here, for explaining the temper brittleness in the amorphous alloys it is possible to explain without contradiction all the experimental facts which do not fit the segregation or crystalline model. In particular, in [4] when examining the effect of different metalloids on the values of T_{br} of the amorphous iron-based alloys, it was noted that the Si–B–C system is not governed by the general tendencies typical of the effect on T_{br} of metalloids such as phosphorus and silicon. The addition of any amount of carbon instead of boron increases T_{br}. A similar difference in the effect of phosphorus and silicon on T_{br}, on the one hand, and of carbon on the other hand, is understandable if it is taken into account that the effect of these elements on cohesion in the iron-based alloys is opposite: phosphorus and silicon cause a rapid decrease whereas carbon increases cohesion.

In addition to this, in a number of investigations (for example [12]) the authors noted a non-monotonic effect of elements such as phosphorus and silicon on T_{br}: the addition of a small amount of these elements (up to 1–3 at.%) may slightly increase T_{br}. However, a further increase of the content of the elements in the alloys results in a decrease of T_{br}. Naturally, not knowing the production parameters, it is difficult to propose a quantitative explanation of the effect of metalloids. Nevertheless, taking into account the proposed model, it may be concluded that up to a certain concentration of phosphorus or silicon in the amorphous alloy the effect of these elements is mainly on the size distribution of the regions of the free volume (displacement to the area of smaller dimensions takes place). During structural relaxation, the atoms-metalloids are bonded in complexes with the iron atoms or trapped in traps – the free volume regions of the minimum size (up to several angstroms). Consequently, they cannot take part in the low-temperature diffusion process which forms the inflow of the atoms decreasing the cohesion the front of the main crack. In all likelihood, the phenomenon of capture of the relatively small phosphorus atoms in traps is associated with the experimentally detected (considerably smaller than could be expected) embrittling effect of phosphorus in comparison with the larger atoms of antimony, tin and cerium. With increase of the phosphorus or silicon content their negative effect becomes stronger. This effect is expressed in the rapidly increasing ductility loss because not all the atoms-metalloids present in the alloy can be any longer bonded in complexes or trapped by traps.

It is also interesting to use the relaxation model of the free volume to investigate the results obtained in [33] for the embrittlement of metalloid-free amorphous alloys which cannot be explained by the segregation model. The ratios T_{cr}/T_{br} for a number of amorphous alloys of different systems were obtained in this study. For the $Cr_{60}Zr_{40}$ and $Ni_{60}Nb_{40}$ alloys this ratio was respectively 1.0 and 0.88. From the viewpoint of the crystalline model of temper brittleness it is not possible to explain these data if it is taken into account that the components of the alloys are 'neighbours' in the periodic system of elements and their electronic concentration are similar. However, the following may be assumed from the viewpoint of the relaxation model of the free volume: on the one hand, a completely different size distribution of the free volume areas is 'quenched' in these alloys, and, on the other hand – the critical size of the micropores determines the possibility of crack opening displacement, may also greatly differ for these alloys.

Attention should be given to the fact that the process of reaching the critical size by the submicropores during structural relaxation may prove to be non-linear because it may consist of two inter-related processes. Recently, a number of studies have appeared [34, 35] linking theoretically and by experiments the low-temperature migration of the free volume areas with local changes of the compositional short-range order of the multicomponent amorphous alloys. The migration of the free volume regions appears to 'break up' the amorphous matrix, reducing the energy barriers for local displacements of the atoms of different type. Consequently, the micropores acting as sinks for the free volume regions during structural relaxation become surrounded by the local interlayer with a considerably higher degree of the compositional short-range order in comparison with the surrounding areas. This in turn results in a decrease of the susceptibility to plastic deformation of this local area and, consequently, increase of the susceptibility to brittle fracture. The final result of the interaction of the migration processes of the free volume regions and the changes of the compositional short-range order should be a decrease of the effective critical size of the submicropores required for the formation of quasi-brittle cracks.

Thus, the main mechanism of temper brittleness is the relaxation mechanism which, however, contains the elements of segregation and crystalline models.

4.6. Inhomogeneities of the structure, forming elastic stress fields under acting loading

Attention will be given to the first stage of formation of the ductile–brittle transition which examines the sources of local stress concentration and the appropriate stress concentration factors both within the framework of the classic (linear) and non-linear (moment) elasticity theory. It is also planned to investigate plastic zones in the region with the maximum stress concentration and areas with the most probable location of the microcracks, and also investigate processes of micro- and macrofracture.

To determine the stress concentration factor with the source of the stresses being the previously mentioned inhomogeneities of the structure of the amorphous alloys, it is necessary to examine three independent problems, namely:

1. Tensile loading of a sheet which, from the physical viewpoint, is an infinite plane with an ellipsoidal pore;

2. Tensile loading of a sheet with an absolutely rigid inclusion; the form of the inclusion in a general case is close to ellipsoidal;

3. Tensile loading of a sheet with an inclusion whose elastic modulus differs from the moduli of the amorphous matrix.

These problems are already investigated below in the framework of both the classic (linear) and moment elasticity theory. It should be mentioned that to evaluate the maximum value of the stress concentration factor k it is sufficient to examine the appropriate planar two-dimensional problems.

The solution of these problems in the classic theory of elasticity assumes the solution of the biharmonic equation with the appropriate boundary conditions for determining the Airy function which in the two-dimensional case, as shown in [36], can be presented in the form of a sum of the potentials $\varphi(z)$ and $\chi(z)$. The determination of the stress state in the problems 1–3 in the moment theory of elasticity requires integrating a system of two differential equations (for the plane stress state) with the appropriate boundary conditions to find to two stress functions $\varphi(x_1, x_2)$ and $\psi(x_1, x_2)$, [37]. The function $\varphi(x_1, x_2)$ should be the solution of the Laplace equation, function ψ is found from solution of the equation $\Delta\psi - l^2 \Delta\Delta\psi = 0$. The latter equation shows that in the moment theory of elasticity there is a new material constant associated with its microstructure and determined as $l^2 = \eta/G$, where η is the ductility, G is the shear modulus. It is well-known that the ductility of the material of the amorphous alloys is a function of temperature T and, therefore, the dependence $l(T)$ must be taken into account to describe the ductile–brittle transition in the amorphous alloys.

The stress concentration factor in a sheet with an elliptical orifice subjected to uniaxial tension. Attention will be given to an infinite sheet under the effect of uniaxial tensile stress p acting in the direction forming the angle β with the X axis (Fig. 4.18). This homogeneous stress state is perturbed by an elliptical orifice, with the main axis of the orifice coinciding with the X axis. The solution has the simplest form in the elliptical coordinates ζ, η [38]:

$$z = c\,\mathrm{ch}\zeta, \quad \zeta = \xi + i\eta, \tag{4.8}$$

where $c = (a + b)/2$, a and b are the half-axis of the ellipse; from which $x = c\,\mathrm{ch}\xi\cos\eta$, $y = c\,\mathrm{sh}\xi\sin\eta$.

The components of the stress tensor in the polar coordinates $X'OY'$ (4.18), obtained by rotating the initial coordinates XOY through the angle β, when the axis OX' becomes parallel to the direction of tensile loading p, has the form

$$\sigma_{x'} + \sigma_{y'} = \sigma_x + \sigma_y,$$
$$\sigma_{y'} + \sigma_{x'} + 2\tau_{x'y'} = e^{2i\beta}\left(\sigma_y - \sigma_x + 2i\tau_{xy}\right). \tag{4.9}$$

The conditions at infinity have the form

$$\sigma_{x'} = p, \, \sigma_{y'} = 0, \, \tau_{x'y'} = 0, \tag{4.10}$$

and, therefore, for the complex coordinates $\varphi(z)$ and $\chi(z)$ of the biharmonic Airy function:

$$4\,\mathrm{Re}\,\varphi' = p,$$
$$2\left[z\varphi'(z) + \chi'(z)\right] = -p\exp(-2i\beta). \tag{4.11}$$

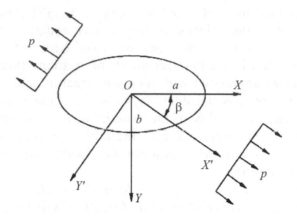

Fig. 4.18. Schematic illustration of the problem of the uniaxial tensile loading of an amorphous matrix containing an elliptical orifice (inclusion).

The boundary of the orifice $\xi = \xi_0$ should be free from stresses having the radial component

$$\sigma_\xi = \tau_{\xi\eta} = 0. \qquad (4.12)$$

All the boundary conditions (4.10)–(4.0) are satisfied if $\varphi(z)$ and $\chi(z)$ have the form [38]

$$4\varphi(z) = Ac\,\mathrm{ch}\zeta + Bc\,\mathrm{sh}\zeta$$
$$4\chi(z) = Cc\,\mathrm{ch}\zeta + Dc^2\mathrm{ch}2\zeta + Ec^2\,\mathrm{sh}\zeta, \qquad (4.13)$$

where A, B, C, D, E are the constants to be determined. Omitting intermediate transformations whose aim is to determine these constants for the boundary conditions, the final expressions for the complex potentials have the form

$$4\varphi(z) = pc\left[\exp 2\xi_0\cos 2\beta\,\mathrm{ch}\zeta + \left(1 - \exp\left(2\xi_0 + 2i\beta\right)\right)\mathrm{sh}\zeta\right],$$
$$\chi(z) = -pc^2\left[\left(\mathrm{ch}2\xi_0 - \cos 2\beta\right)\zeta + \exp 2\xi_0\mathrm{ch}2\left(\zeta - \xi_0 - i\beta\right)/2\right], \quad (4.14)$$

where the subscript denotes the boundary of the orifice.

The circumferential component of the stress σ_η at the boundary of the orifice is determined from the relationships linking the components of the stresses in the elliptical and rectangular coordinates (4.12), (4.13) and the first of the equations of the system (4.11):

$$\sigma_\xi + \sigma_\eta = \sigma_x + \sigma_y; \tag{4.15}$$

$$\sigma_\eta - \sigma_\xi + 2\tau_{\xi\eta} = y^{2i\beta}\left(\sigma_y - \sigma_x + 2i\tau_{xy}\right); \tag{4.16}$$

$$\left(\sigma_\eta\right)_{\xi=\xi_0} = p\left[\operatorname{sh} 2\xi_0 + \cos 2\beta - \exp\left(2\xi_0\right)\cos\left(\beta-\eta\right)\right]\left[\operatorname{ch} 2\xi_0 - \cos 2\eta\right]^{-1}. \tag{4.17}$$

When the tensile force p acts under the right angle to the main axes XOY ($\beta = \pi/2$) one obtains

$$\left(\sigma_\eta\right)_{\xi=\xi_0} = p\left(\exp 2\xi_0\right)\left[\left[\operatorname{sh} 2\xi_0\left(1+\exp\left(-2\xi_0\right)\right)\right]\left[\operatorname{ch} 2\xi_0 - \cos 2\eta\right]^{-1} - 1\right]. \tag{4.18}$$

This component has the maximum value at the ends of the major axis ($\cos 2\eta = 1$):

$$\left(\sigma_\eta\right)_{\xi=\xi_0, \eta=\pi/2} = p\left(1 + 2a/b\right). \tag{4.19}$$

The stress concentration factor k increases without bounds with increase of the ratio between the major and minor axes of the ellipse. When $a = b$, then $k = 3$ which coincides with the value obtained for the circular orifice. The minimum value $k = -1$ is obtained at the ends of the minor axis of the elliptical orifice. If the tensile stress p is parallel to the major axis ($\beta = 0$), the maximum value of k at the circumference is obtained at the end of the minor axis $k = 1+2b/a$. When the width of the elliptical orifice is reduced, k tends to 1. At the ends of the major axis $k = -1$ for any values of a/b.

It is important to mention the results obtained in [39] for the pure bending of a sheet in which the elliptical orifice was at equal distances from its surfaces. The maximum value of k is as follows:

$$k = 2Ra/b, \quad R = (a + b)/2. \tag{4.20}$$

The stress concentration factor for the circular orifice in the moment theory of elasticity is

$$k = (\sigma_{\theta\theta})_{\max}/p = (3 + F)/(1 + F), \tag{4.21}$$

where $F = 8(1 - v)/[4 + a^2/l^2 + 2a/l + K_0(a/l)]$, a is the radius of the orifice, K_0 is the modified Bessel function of the second kind and zeroth degree, K_1 is the modified Bessel function of the second kind and first degree. If $F = 0$, k is equal to a constant value obtained by the classic theory of elasticity, namely $k = 3$. This case forms when there are no moment stresses. The characteristic constant of the material l tends to 0 because

$$\lim_{a/l \to \infty} \left[K_0(a/l) / K_1(a/l) \right] = 1.$$

However, if the moment stresses differ from zero and $l \neq 0$, k depends on the Poisson coefficient and the ratio of the radius of the orifice to the characteristic constant of the material l. At $l \to \infty$, i.e. when $a/l \to 0$, the value k differ from $k = 3$, and the deviation from the classic value increases with a decrease of a/l. Consequently, taking into account the moment stresses reducing k, and the 'softening' effect of k increases with decreasing l. The minimum values of k equal: $k = 2.4$ ($v = 0.5$) and $k = 2.6$ ($v = 0$), i.e., decrease by 13–20% for the value $a/l = 3$.

The stress concentration factor in a sheet with an absolutely rigid inclusion subjected to uniaxial tensile loading. The formulation of this problem is identical with that of the problem examined previously [40]. The result of the effect of tensile stress p, forming the angle β with the major axis a of the elliptical inclusion (axis OX), – the translational movement and the rotation of the latter. The translational movement may be ignored because it can be eliminated by displacing the entire system. This shows that it is necessary to determine the components of the stress sensor of the material of the sheet caused by rotating the inclusion through the angle ε. Since the inclusion is absolutely rigid, the contour values of the component of the displacement vector are equal to:

$$g_1 = -\varepsilon y, \quad g_2 = -\varepsilon x. \tag{4.22}$$

This problem will be solved by the conformal imaging method. The function carrying out imaging of the plane with the elliptical inclusion on the plane with the circular inclusion has the form [36]

$$z = \omega(\xi) = R(\xi + m\xi^{-1}), \tag{4.23}$$

where $R > 0$, $0 \leq m < 1$, $a = R(1 + m)$, $b = R(1 - m)$. Selecting the parameters R and m it is possible to select parameters for the elliptical inclusion of any size and shape. The contour values of the components of the displacement vector and its complexly conjugated values are:

$$g = i\varepsilon(x+iy) = i\varepsilon z = i\varepsilon R(\sigma + m\sigma^{-1}), \quad \overline{g} = -i\varepsilon R(\sigma^{-1} + m\sigma). \quad (4.24)$$

Calculating the Cauchy integrals from $g(\sigma)$ and $\overline{g}(\sigma)$ for the contour of the inclusion and assuming $\psi_0(\infty) = 0$ gives the final expressions for the complex potentials (it should be noted that $\psi(z) = \chi'(z)$):

$$\int_T h(T_a, T, t)dT = 0,$$

$$\int_T h(T_a, T, t)dT / \int_T h(T_a = 293\,K, T, 0)dT, \quad (4.25)$$

where $\kappa = (3 - v)/(1 + v)$, v is the Poisson coefficient, G is the shear modulus. The angle of rotation ε is determined from the condition of equality to 0 of the main momentum of the forces M_0 acting on the inclusion from the side of the surrounding material. Momentum M_0 is determined by the increment of the part of the complex potential

$$\chi_1(z) = \int_\sigma \Psi_1(z)dz = \int_\sigma \Psi_1(\xi)\omega'(\xi)d\xi. \quad (4.26)$$

As a result of the integration of (4.26) and subsequent dividing into the real and imaginary parts one obtains the expression for the momentum M_0

$$M_0 = 4\pi G\varepsilon R^2 (1 + m^2/\kappa) - 2\pi m R^2 C'(1 + \kappa^{-1}). \quad (4.27)$$

The condition $M_0 = 0$ gives the angle of rotation of the inclusions ε:

$$\varepsilon = pm(1+\kappa)\sin 2\beta / (4G[m^2 + \kappa]). \quad (4.28)$$

The value of the radial component of the stress tensor σ_p has the maximum value and, consequently, the form of the stress concentration factor $k(\theta)$ at the contour of the inclusion is

$$k(\theta) = \left[(a\sin\theta)^2 + (b\cos\theta)^2\right]^{-1} \times \left\{(b\cos\theta)^2 + \left[a(1-v) - 2bv\right]a(\sin\theta)^2 + \right.$$
$$\left. + \left[\left[2a - vb(1-v)\right]\left[b(\cos\theta)^2\right]\right]/\left[3 + 2v - v^2\right]\right\}. \qquad (4.29)$$

The maximum value of k is obtained at $\beta = \pi/4$ and $\theta = 0$.

The stress concentration coefficient in a sheet with a cylindrical inclusion under uniaxial tensile loading. The determination of the stress state in the sheet, containing an inclusion of an arbitrary, especially elliptical shape with the elastic moduli different from the moduli of the matrix, is associated with considerable computing difficulties [41] caused by the need to solve the conjugation problem. However, if the shape of the inclusion differs only slightly from the circular form, it is possible to describe quite accurately the effect of the difference in the elastic moduli of the inclusions in the matrix on the variation of the pattern of the stress state and, consequently, the stress intensity factor. It is therefore possible to avoid cumbersome computations, also taking into account the fact that the final expressions are very complicated and difficult to understand, if the stress states in the sheet for two limiting cases: the orifice and the absolutely rigid inclusion of the given shape, are known.

The effect of the difference in the elastic moduli of the inclusion material and the sheet material on the stress concentration factor will be shown on the example of a cylindrical inclusion, comparing the results obtained by the methods of classic (linear) elasticity theory and the theory of momentum stresses. The shear modulus, the Poisson coefficient and the characteristic length (a new material constant developed in the theory of the momentum stresses) will be denoted by the symbols G_1, v_1 and l_1 for the material of the inclusion and G_2, v_2, and l_2 for the material of the matrix. The solution procedure is as follows. The equilibrium equations of the momentum theory of elasticity are solved for the potentials φ, ψ, χ in the polar coordinate system for the material of the inclusion and the matrix. The boundary conditions are the continuity on the interface of the radial, tangential and momentum components of the stress tensor, the displacement vector and the angle of rotation. Analytical expressions for the components of the stress tensor in the material of the sheet and the inclusion are very complicated, cumbersome and difficult to understand. The following conclusions can be made on the basis of the calculations.

If $g = G_2/G_1 > 1$, the stress concentration forms in the material of the sheet, if $g < 1$ it forms in the inclusion material. The variation of the stress concentration factor k depends on the values of the parameters $\lambda_1 = a/l_1$ and $\lambda_2 = a/l_2$, where λ_i is the argument of the modified Bessel function of the second kind and the second degree; if the inclusion is 'soft' ($g > 1$), k is bounded at the top; if the inclusion is 'hard' ($g < 1$) it is bounded at the bottom by the values following from the classic linear theory of elasticity. Numerical analysis was carried out for the cases when $v_1 = 0.25$ and $v_2 = 1/3$, $g = 2$ and $g = 1/2$ for different values of the arguments $\lambda_1 = a/l_1$ and $\lambda_2 = a/l_2$. For the 'soft' inclusion when $g = 2$, k increases with increase of the ratio $\lambda_2 = a/l_2$, and in the limit ($\lambda_2 \to \infty$) coincides with the value following from the classic theory of elasticity ($k \approx 1.58$). If $g = 1/2$ ('rigid' inclusion), k increases with a decrease of the ratio $\lambda_2 = a/l_2$ and has the maximum value $k \approx 1.4$. Thus, the effect of momentum stresses on the stress concentration factor is completely opposite to the case of tensile loading the sheet with an orifice. In comparison with the results of the classic theory of elasticity in the case of the 'rigid' inclusion the value of k tends to increase.

It is interesting to examine a special case in which the inclusion is absolutely 'rigid'. The components of the stress tensor in the sheet at the junction with the inclusion are equal to:

$$\left[\sigma_{\theta\theta}(a,\theta)\right]_2 / p = v_2\left[1 + 2\cos 2\theta / (3 - 4v_2 - F_2)\right]; \qquad (4.30)$$

$$\left[\sigma_{rr}(a,\theta)\right]_2 / p = (1 - v_2)\left[1 + 2\cos 2\theta / (3 - 4v_2 - F_2)\right]; \qquad (4.31)$$

$$F_2 = 4(1 - v_2)K_1(\lambda_1) / \lambda_2 K_2(\lambda_2), \qquad (4.32)$$

where K_1 is the modified Bessel function of the second kind and first degree, K_2 is the modified Bessel function of the second kind and second degree, $\lambda_2 = a/l_2$, where a is the radius of the inclusion, l_2 is the characteristic length of the material of the amorphous matrix. Since $0 \leq v_2 < 0.5$, then when the inclusion is absolutely rigid, σ_{rr} is always greater than $\sigma_{\theta\theta}$, and both components have the maximum values at $\theta = 0$. In this case

$$k = (1 - v_2)\left[1 + 2 / (3 - 4v_2 - F_2)\right], \qquad (4.33)$$

as shown in Fig. 4.19.

The solution of the problems (4.31)–(4.33) can be used to determine in the analytical form the stress concentration factor for uniaxial tensile loading of the amorphous matrix containing an ellipsoidal orifice, an absolutely rigid ellipsoidal inclusion and evaluate the values of k when the elastic moduli of the matrix and the inclusion differ. Above all, it should be noted that the maximum stress concentration forms in the uniaxial tensile loading of the matrix containing the ellipsoidal orifice and the absolutely rigid inclusion. Comparison of the results obtained in the framework of the classic and momentum theories of elasticity indicates that taking into account the momentum stresses slightly decreases (by 15–20%) the stress concentration in the amorphous matrix for the orifice whose contour is almost circular. For the inclusion with the elastic moduli different from the moduli of the matrix, the stresses are bounded at the top, when the inclusion is 'soft' ($g > 1$) and at the bottom when the inclusion is absolutely 'rigid' ($g < 1$), using the values following from the solution of the appropriate problem

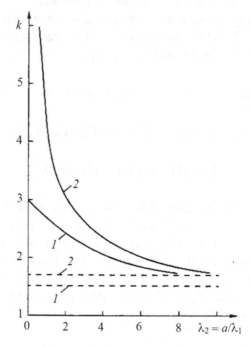

Fig. 4.19. Dependence of the stress concentration factor k in uniaxial tensile loading of the amorphous matrix containing an absolutely rigid inclusion: solid lines indicate k calculated from the momentum theory of elasticity, the dotted lines – from the classic theory; curve 1 – the Poisson coefficient $\nu = 1/2$; curve 2 – $\nu = 0$.

by the methods of the classic theory of elasticity. This consideration shows that the most powerful stress raisers are the pores and the absolutely rigid inclusions. In fact, the last case is realised when the elastic moduli of the amorphous matrix and the inclusion differ by no more than an order of magnitude. In addition, this case is especially important, as indicated by the momentum theory of elasticity, if the new constant of the material $l = \eta/G$ is comparable with one of the linear dimensions of the inclusion because the value of the stress concentration factor k in this case, as shown by Fig. 4.19, may be very high. In addition to this, it is important to take into account the temperature dependence of the new material constant l formed in the momentum theory of elasticity and varying both during heating of the amorphous alloy to temperature T_a and in subsequent cooling to room temperature.

Since to determine the stress concentration factor it was necessary to obtain in the analytical form all the components of the tensor of elastic stresses in the amorphous matrix containing pores and inclusions and subjected to uniaxial tensile loading, it is possible to transfer directly to the next stage of the general problem and, in particular, the solution of elastoplastic problems for the most important cases: the orifice and the absolutely rigid ellipsoidal inclusion.

4.7. Acoustic emission studies

The acoustic emission method can be used to investigate micro- and macroscopic processes of both plastic deformation and failure of the amorphous alloys [42]. The experiments were carried out on specimens of amorphous ribbons based on iron and cobalt in the tests of the latter by uniaxial tension and bending in both the initial condition and after annealing with the parameters $T = 593$ K, $\tau = 1$ h. In the latter case, the amorphous alloy was in the embrittled state [16]. The dimensions of the specimens were $0.02 \times 15 \times 50$ mm. The multilevel analysis method was used which is suitable for determining the statistical characteristics of the pulsed flows, for unfolding individual pulses in time and for spectral analysis of the structure. The duration of the pulsed flows corresponded to the time period of loading the ribbon specimens and equalled 40 s in bending the specimens in the initial condition, and in the case of the embrittled specimens to complete fracture: $\tau \approx 12$ s (Fig. 4.20*a*).

The freshly quenched material of the amorphous alloy was tested
to obtain and analyse the pulsed acoustic emission (AE) fluxes,
unfold the individual pulses in time and investigate their internal
structure; in addition to this, formation of the shear bands was
studied by scanning electron microscopy (SEM). Regardless of the
dependence on the type of mechanical tests, the plastic deformation
of the amorphous alloys was characterised by the presence of
the shear bands with adjacent steps. In the quantitative aspect of
the uniaxial tension and bend tests differed: in the first case, no
more than one or two shear bands formed, in the second case their
number increased by two or three orders of magnitude. In the time
dependences of the AE fluxes the individual pulses appeared at the
start of plastic flow, the average time between the individual pulses
was approximately 250 µs. The time–pulse analysis shows that the
distribution of the intervals between the individual pulses corresponds
to the Poisson statistics $\chi^2 = 1.08$; $\chi_c^2 = 9.49$; $\chi^2 < \chi_c^2$ is the criterion

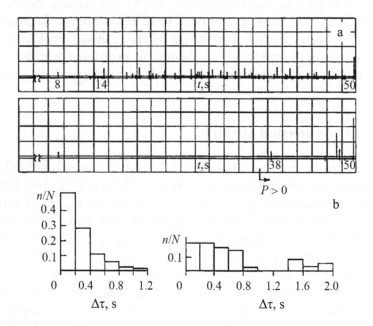

Fig. 4.20. Time dependences of the pulsed acoustic emission fluxes obtained in
bend tests of the ribbon specimens of the amorphous alloy; the upper graph is the
time dependence of the pulsed AE fluxes of the specimens of the freshly quenched
amorphous alloy; the lower graph – annealed at $T_a = 593$ K, $\tau = 1$ h (a). The statistical
distribution of the intervals between the AE pulses of the freshly quenched specimens
of the amorphous alloy (a): the left graph corresponds to the Poisson distribution
of the intervals for the freshly quenched ribbon; the right graph to the correlated
distribution after the ductile–brittle transition (b).

of agreement with the Poisson distribution law, χ_c^2 is a critical value at the significance level of $c = 0.05$ (Fig. 4.20b). Analysis of the oscillograms shows that the duration τ of the pulses of the same amplitude varied in the range 400–500 μs (the average value $\tau = 450$ μs). The comparison of the number of the shear bands using the SEM results with a number of passes showed that they are in quantitative agreement. This shows that each pulse resulted in the formation of a single shear band, and the duration of the pulse corresponded to the duration of formation of the shear band.

The characteristics of the individual AE pulses, accompanying the formation of shear bands, will be discussed again. Their spectral analysis was carried out by multiple reproduction from a video recorder (in the 'stop frame' mode) to a spectral analyzer and a graph plotter. This was followed by determining the pulse parameters W and α. The first of these parameters – the pulse energy – was obtained by the numerical integration of the spectral density function $\int G(f)df$ (in relative units) by the trapezoidal method in the frequency range 20–1000 kHz, the second – coefficient α, characterising the form of the spectrum – was determined as the ratio of the energy, relating to the frequency range 150–1000 kHz, to the total pulse energy. The values of W and α were averaged out throughout the entire ensemble of the pulses of the same time.

The pulses with the truncated envelope represented ≈70% of the total number of the pulses observed in deformation of the amorphous ribbons. The remaining pulses were characterised by continuous filling and differed by the small spectral energy, with the average value of $\overline{W} \approx 6.4$ at $\alpha = 0.5$. The pulses with the complicated substructure had the parameters $\overline{W} \approx 12.3$ $\alpha = 0.52$. It may be assumed that the pulses with the continuous filling corresponded to the process of local strengthening in the shear band. In the shear pulses with the truncated envelope experiments were carried out to define one filling frequency $f_1 \approx 30$ kHz, and in some two frequencies: $f_1 \approx 30$ kHz and $f_2 \approx 500$ kHz.

The difference between the AE of the specimens after the ductile–brittle transition and in the condition after quenching is shown clearly by comparing Fig. 4.20a and b. In the case of the embrittled specimens it is necessary to talk about the microplasticity of the individual sections because the total number of the shear pulses in these quenched specimens was an order of magnitude smaller than in the freshly specimens. The time–pulse analysis showed that in the annealed specimens of the amorphous alloy the flux of the shear pulses was ordered in accordance with the χ-criterion: $\chi^2 > 50$; $\chi_c^2 = 43.8$; $\chi^2 > \chi_c^2$. The duration of the pulses of the same amplitude varied in the range $\tau = 400$–600 µs.

The main difference between the pulses accompanying the fracture of the freshly quenched (in the uniaxial tensile test) specimens and the specimens after the ductile–brittle transition (in the uniaxial tension and bending test) of the ribbon specimens of the amorphous alloys was the presence of 2–3 powerful pulses whose energy was 2–3 times higher than the energy of the pulses formed at the moment of fracture of the freshly quenched specimens. In addition to this, the specimens accompanying the processes of crack formation and macrofracture of the annealed specimens of the amorphous alloys were characterised by continuous filling without the internal substructure.

The SEM data for the width of the steps adjacent to the shear bands, and the average duration of the shear pulses (according to the results of time-pulse analysis) were used to determine the average speed of movement of the shear bands equal to $7.8 \cdot 10^{-5}$ m/s which completely coincides with the lower estimate of this value obtained by high-speed filming ($4 \cdot 10^{-5}$ m/s [17]).

It should be noted that the freshly quenched specimens of the amorphous alloy were characterised by shear pulses without internal filling (20–30% of the total number) which are characteristic of the strain hardening process [43]. The presence of the Kaiser effect, detected in the investigated amorphous alloys, confirms the effect of strain hardening and explains the existence of the previously mentioned pulses.

It should be stressed that the non-Poisson nature of the pulsed statistics in the specimens of the amorphous alloys subjected to the ductile–brittle transition is in complete agreement with the data in [44, 45] where is shown that the correlated AE flux corresponds to brittle fracture.

4.8. Formation of embrittled surface layers

The authors of the book carried out detailed systematic studies of the formation of embrittled surface layers (ESL) for amorphous alloys of the metal–metalloid type of different compositions [46]. The experimental results show that when the ESL reaches the critical thickness of ~2 µm on both the contact and free surfaces of the ribbon after the temperature–time effects of specific temperature and duration, the samples of the amorphous alloys start to fail in the bend test [40]. The critical thickness of the ESL has the simple physical meaning. It is equal to the minimum length of the brittle crack, equal to the thickness of the ESL, causing fracture of the specimens of the amorphous alloy in the given type of mechanical testing. The mechanism of nucleation of the cracks in the ESL is discussed in the second part of the article. Comparison of the evolution of RS and the mechanical behaviour of the specimens of the amorphous alloys in the bend test was carried out to formulate the criterion for the start of the ductile–brittle transition [16]

$$\int_T h(T_a, T, t)\, dT = 0, \tag{4.34}$$

where T_a is the isochronous annealing temperature, t is its duration, $h(T_a, T, t)$ is the spectral density of heat generation, the lower integration limit corresponds to the start of heat release, the upper limit to the temperature corresponding to the point at the half width of the left branch of the high-temperature part of the RS (Fig. 4.21 [16]). The ductile–brittle transition criterion (4.34) was verified by experiments for the amorphous alloy of the metal–metalloid type with more than 25 different compositions. The criterion was used as a basis for developing a method of predicting the state of the material of the amorphous alloy confirmed by the results of mechanical tests on the specimens of the amorphous alloy both after holding for 8000 hours at room temperature and in the course of low-temperature ageing at $T = 455$ K for 100 hours. The results indicate that the macroscopic processes causing the formation of the ESL in both high-temperature annealing and low-temperature ageing and also in holding the amorphous alloy in the range of climatological temperatures are identical.

Further studies use the following procedure. The amorphous alloy of the $Fe_{61}Co_{20}B_{14}Si_5$ composition was subjected to embrittling annealing causing the formation of an ESL with a thickness of 2 μm on the contact and free surfaces of the ribbon specimens. After removing the ESL by some method (chemical etching, electrolytic polishing, machining, etc) the ribbon specimens were subjected to mechanical tests by 180° free bending. After the tests, the specimens showed the residual bend angle similar to the angle obtained for the freshly quenched ribbon specimens. The same specimens were then annealed (second annealing) with the same parameters. After the thermal–time effects, the ESL of the same thickness (~2 μm)

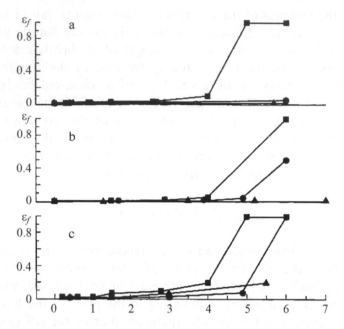

Fig. 4.21. Dependence of the relative fracture strain ε_f as a function of the thickness of the layer removed by etching Δ from the ribbon specimens of the amorphous alloy after preliminary embrittlement: a – $Fe_{70}Ni_8B_{13}Si_9$, b – $Co_{70}Fe_6Si_{15}B_{10}$, c – $Co_{74}Fe_3Mn_3B_{15}Si_5$ as a result of preliminary annealing with the parameters $T_a = 553$ (a), 643 (b) and 593 K (c) with the duration, h: ■ – 1, ● – 2, ▲ – 6.

was again produced and the subsequent bend test of the specimen caused them to fail. After removing the ESL the specimens were plastically deformed like the freshly quenched specimens. Similar results were obtained after the third and fourth annealing cycle. In addition to this, differential scanning calorimetry (DSC) was applied to obtain relaxation spectra of the specimens with the ESL and also after removing the ESL. In the specimens with the ESL the normalised heat generation of the low-temperature part of the RS $\int h(T_a,T,t)dT / \int h(T_a = 293 \text{ K}, T,0)dT$ was a negative value, the specimens from which the ESL was removed showed positive values [16, 47] on the basis of these investigations in may be concluded that the ductile – brittle transition of the amorphous alloy of the metal–metalloid type stars with the formation of the ESL which are not capable of plastic shape changes.

Investigations of the effect of the annealing parameters (temperature and duration) on the ESL thickness (in addition to the amorphous alloys of the composition $Fe_{61}Co_{20}B_{14}Si_5$) were also carried out for the amorphous alloys of other compositions, namely $Fe_{70}Ni_8B_{12}Si_9$, $Co_{70}Fe_6Si_{15}B_{10}$ and $Co_{74}Fe_3Mn_3B_{15}Si_5$. As in the $Fe_{61}Co_{20}B_{14}Si_5$ amorphous alloy, the ductile–brittle transition in the previously mentioned alloys started with the formation of the ESL, and the ductility was restored as a result of the removal of the ESL by chemical etching in a 5% solution of nitric acid. The thickness of the removed layer was determined by weighing. Since the etching process was non-uniform, each point in the $\varepsilon_f - \Delta$ coordinates was obtained by averaging for 30 specimens, where ε_f is the relative fracture strain calculated from the relation $\varepsilon_f = (d - \Delta)/(D - (d - \Delta))$, Δ is the thickness of the removed layer, d is the thickness of the ribbon, D is the distance between the loading plates at the moment of fracture in the bend testing of the ribbon specimens (Fig. 4.21).

Figure 4.21 shows that the rate of growth of the ESL whose thickness determines the relative fracture strain in the mechanical tests of the ribbon specimens of the amorphous alloys, depends only slightly on the composition of the amorphous alloy and the controlling role is played by the temperature and duration of preliminary annealing. At the duration of isochronous annealing equal to 1 h, the investigated amorphous alloys showed the restoration of the initial ductility; after $t_a = 2$ h, only the amorphous alloy (a) and partially (b) showed complete restoration of ductility; annealing for six hours resulted in the complete transition of the entire material of the amorphous alloy (a–c) to the brittle state. The highest stability with respect to the thermal effects was obtained for the $Co_{74}Fe_3Mn_3B_{15}Si_5$ alloy due in most cases to the presence of manganese in its composition. On the basis of Fig. 4.21a–c it may be concluded that the transition of the amorphous alloy from the ductile to brittle state is associated with the formation of the ESL, and the increase of the thickness of the ESL after the maximum six hour annealing resulted in the situation in which it was not possible to restore the initial ductility by removing the ESL which occupied the entire volume of the amorphous ribbons. It should be mentioned that the additional investigations of the amorphous alloys, carried out by differential scanning calorimetry, shows radical changes in the relaxation spectrum based on the complete disappearance of its low-temperature part, with the exception of the cases of restoration of initial ductility.

Thus, the phenomenon of temper brittleness is quite complicated. This is associated with the simultaneous occurrence of the processes of structural relaxation, diffusion, the redistribution of the free volume and chemical components of the amorphous alloy, with the surfaces of ribbons acting as sinks for these components in the initial stages of the ductile–brittle transition.

4.9. Investigation of the magnetic, magneto–optical and spectro-ellipsometric properties

The magnetic and magneto–optical properties of the amorphous alloys of the composition $Fe_{61}Co_{20}B_{14}Si_5$ were investigated. The ribbon specimens of the alloys were subjected to isochronous annealing with the duration $t_a = 10$ min at temperatures $T_a = 623$, 648 and 673 K, corresponding to the temperature of the start of the ductile–brittle transition [48]. As regards the results of the magnetic studies, the most characteristic are the dependences of the angular delay of the magnetic moment on annealing temperature T_a and the strength of the external magnetic field. The field dependence of the hysteresis of the angular delay of the magnetic moment in a rotating magnetic field indicates the formation of a highly anisotropic phase in the specimens annealed at $T_a = 623$ K with a maximum in the field with the strength $H \approx 17$ Oe. In the specimens annealed at $T_a = 648$ K, the maximum is displaced to the range of the fields with higher strength and disappears when the T_a of the specimens equals 673 K. Consequently, it may be concluded that at $T_a = 623$ K the ESL material starts to separate into two amorphous phases, depleted and enriched with the metalloid atoms. The new amorphous phase, enriched with the atoms of the metal, is highly anisotropic and its amount is small. When the T_a is increased to 640 K the size of the phase rapidly increases because the anisotropy increases, and at $Ta = 673$ K the specimens again become structurally homogeneous because of the decrease of the anisotropy as a result of the joining of the highly anisotropic amorphous phase [49, 50]. Similar processes also take place in cases in which the ESL overlaps the entire cross-section of the specimen. In the latter case, the magnetisation of the amorphous alloy is increased, and areas enriched and depleted in the metalloid atoms appear in the volume of the material.

Magneto–optical investigation of the equatorial Kerr effect (EKE) fully confirmed the results of magnetic studies. The EKE slightly decreases for the specimens of the amorphous alloys annealed at $T_a = 623$ K in comparison with the EKE of the freshly quenched specimens. However, in the case of the specimens annealed at $T_a = 614$ K the EKE increases by more than 60% of the minimum value (at T_a = 623 K). This increase is explained, in particular, by the formation of a new highly anisotropic amorphous phase in which the environment of the iron atoms is close to octahedral, and also by the increase of the magnetisation of the samples [49]. It should be stressed that all the effects described above, associated with the formation during annealing of the regions in the ESL enriched and depleted in the atoms of the metalloids, disappear after the removal of the ESL whose thickness does not exceed 2 μm on both sides of the ribbon of the amorphous alloy.

In [51] the method of angular ellipsometry was used to investigate the change of the optical polarisation characteristics of the ESL formed as a result of annealing in the initial stage of the ductile–brittle transition. The dependences of the main angle of incidence of light φ_p on T_a were obtained for the amorphous alloys of the Si–Co–Ni–Si, Fe–Ni–B and Fe–Co–Si–B systems. The observed decrease of the values of φ_p (and, correspondingly, the optical conductivity) for the annealed amorphous alloys is explained by the formation during annealing of structural inhomogeneities in the ESL, causing an increase of the number of collisions of the conduction electrons with these inhomogeneities. The optical method was used to estimate the thickness of the ESL ($\approx$2 μm) containing structural micro-inhomogeneities, and the increase of the thickness of the ESL with increasing T_a was confirmed.

The results of investigation of the physical properties in the initial stage of the ductile–brittle transition can be used to draw the following conclusions:

- the physical properties of the ESL material, located within the limits of the amorphous state, differ from the properties of the initial amorphous matrix;
- the ESL contains areas enriched and depleted in the atoms of the metalloids; the new amorphous highly isotropic magnetic phase, depleted in the atoms of the metalloids, overlaps the entire volume of the ESL with increasing T_a;
- the thickness of the ESL is a function of the annealing parameters; increase of annealing temperature T_a or annealing time t_a results in overlapping of the ESL of the entire cross-section of the amorphous ribbons. However, the analytical dependences of the growth of the ESL as a function of the annealing parameters have not as yet been defined.

-

4.10. General model of failure in the ductile–brittle transition

In the classic linear theory of elasticity the stress concentration factor k does not depend on the relation between the linear dimensions of the stress raisers (cavities, inclusions, cracks, etc.) and the characteristic constant of the material I used in the momentum theory of elasticity and has the dimension of length. In [37, 40, 41] it is shown that if any linear size of the stress raiser L satisfies the ratio $I \ll L$, then the additional stresses, characterised by the value k, are macroscopic. Otherwise ($I \gg L$) they are local and the stress concentration factors depend on the ratio $\eta = L/I$. In the macroscopic case $k = \sigma^{max}/\sigma$, where σ^{max} are the maximum values of the stresses, σ are the stresses which would act in the absence of the stress raiser. However, if the stresses are local, then in the limiting case when any linear size of the stress raiser satisfies the ratio $L_i \ll I$, where $i = 1, 2,..., n$, the stress concentration can be ignored in the limit. For example, for a circular orifice in the macroscopic case $k = 3$, if the radius of the orifice is $R \ll I$, then, as shown strictly in [41], $k = 1$. In [37] the authors defined more accurate boundaries of the application of the results of momentum and micropolar theories of elasticity. For the momentum elasticity theory they determined the strict restrictions when the results in the quantitative aspect must be taken into account in comparison with the classic linear theory of elasticity. They are expressed by the ratio $0.1 \le L/I \le 10$. The restrictions for the micropolar elasticity theory are even more severe. They are described in detail in [40].

The case postulated in this book will now be discussed. In [41] it was shown that for an elastic plane with a macroscopic slit the critical size of the slit is determined by the relationship

$$\sigma_{cr} = \sigma_0 / k = 2\sigma_0 \left[I/a \right]^{1/2} / \beta_v, \qquad (4.35)$$

where σ_{cr} is the maximum value of the external stress corresponding to the critical size of the crack in the form of a mathematical section; σ_0 is the external stress in the absence of this stress raiser; k is the macroscopic value of the stress concentration factor; β_v is the numerical coefficient equal to $\beta_v = [2(1-2^{1/2}/(1+2^{1/2})^{1/2})v + [3 \cdot 2^{1/2} - 1]/[2(1 + 2^{1/2})^{1/2}]]2^{1/2}$; v is the Poisson coefficient. The resultant relationship coincides with the accuracy to the constant multiplier with the Griffith equation [41] for the critical external load

$$\sigma_{cr} = \left[2E\gamma / \pi \left(1 - v^2\right) a \right]^{1/2}, \tag{4.36}$$

here E is the Young modulus, γ is the specific surface energy of the crack, a is the crack length. Equating the relationships (4.35) and (4.36) leads to the value of the characteristic constant of the material

$$I = \left[\beta_v^2 / 2\pi \left(1 - v^2\right) E\gamma / \sigma_0^2 = \mu E\gamma / \sigma_0^2 \right], \tag{4.37}$$

where μ is equal to the expression enclosed in the square brackets. For the Poisson coefficient $v = 0.3$ the numerical value is $\mu = 0.421$.

The variation of the characteristic constant of the material of the amorphous alloy during the ductile–brittle transition will be determined. Numerical calculations were carried out for an amorphous alloy of the composition $Co_{75.4}Fe_{3.5}Cr_{3.3}Si_{18.7}$. The experimental values determined previously for the fracture toughness K_{1c} were $K_{1c}^{(1)} \approx 45$ MPa $\cdot$ m$^{1/2}$ (at the start of the ductile–brittle transition) and $K_{1c}^{(2)} \approx 10$ MPa $\cdot$ m$^{1/2}$ (after completion of the transition). The values of the Young modulus determined in the tests of annealed ribbons in uniaxial tensile loading corresponding to the start and completion of the examined process are equal to $E^{(1)} = 140$, $E^{(2)} = 170$ GPa. The values of the surface energy per unit length of the brittle crack were calculated from the relationship

$$\gamma = (K_{1c})^2/2E \tag{4.38}$$

and were equal to $\gamma^{(1)} = 7.232 \cdot 10^3$ and $\gamma^{(2)} = 294$ J$\cdot$m^2. It should be noted that the value $\gamma^{(1)}$ was determined without taking into account the scattering of energy by the growing crack with the nucleation of new shear bands and the shear bands which continue to grow. Thus, all the values included in the equation (4.37) are available (numerical values of σ_0 were determined by testing the ribbon specimens of the amorphous alloy in uniaxial tensile loading using the dependence σ–ε). The values of I for the amorphous alloy with the composition $Co_{75.4}Fe_{3.5}Cr_{3.3}Si_{18.7}$ were equal to $I^{(1)} = 2.34 \cdot 10^4 = 234$ μm; $I^{(2)} = 19$ μm. Thus, the material of the amorphous alloy ribbons after annealing, corresponding to the start of the ductile–brittle transition, should be regarded as a homogeneous elastoplastic medium ignoring the presence in it of structural heterogeneities of different nature because their size is considerably smaller than the characteristic size I and the additional stresses caused by them in active loading are local. For the material of the amorphous alloys after completion of the ductile–brittle transition macroscopic heterogeneities are those where one of the linear dimensions is not smaller than $L \geq 1.9$ μm, and the additional fields of the elastic stresses generated by them and characterised by the stress concentration factor k are macroscopic quantities. Similar results were obtained also for the amorphous alloy with the $Fe_{61}Co_{20}B_{14}Si_5$ composition.

A planar problem of the nucleation of a brittle crack in the ESL and also in the amorphous alloy after the ductile–brittle transition will be discussed. The maximum stresses in bending the ribbon specimens form at the vicinity of the ribbon surface and, therefore, the calculation diagram shown in Fig. 4.22 can be used with good approximation. The pore (in the three-dimensional case an orifice) is situated at distance d from the free surface. Since the thickness of the ESL in the initial stage of the ductile–brittle transition is at least 10 times smaller than the thickness of the ribbon, the latter can be regarded as a half plane and the stresses present in the ESL as a constant value because their change at the thickness of the ESL equal to 2 μm and does not exceed 2% for the typical transverse dimensions of the amorphous alloy ribbons, ~25 μm.

A similar problem was solved for the first time by Jeffry [52]. The half plane (amorphous alloy ribbon) was tensile loaded with stresses p = const parallel to the free boundary of the half plane. The components of the stress tensor were determined for mutually perpendicular, bipolar and curvilinear systems of coordinates ξ and η. The stresses $\sigma_\eta = \sigma_{br}$ at the boundary of the half plane were

$$\sigma_{br} = p\left[1 + (1 - \cos\eta)\sum_{n=1}^{\infty}P_n\cos n\eta\right]. \tag{4.39}$$

The stresses at the contour of the circular orifice are determined by the equation

$$\sigma_\eta = 2p\left[1 - \left(2sh^2\xi_1\sin^2\eta\right)/\left(ch\xi_1 - \cos\eta\right)^2 + \left(\cos\xi_1 - \cos\eta\right)\times\right.$$
$$\left.\times\left[\left(2sh\xi_1\right)^{-1} + 2e^{-2\xi_1}\cos\eta + \sum_{n=1}^{\infty}P_n\cos n\eta\right]\right]. \tag{4.40}$$

The parameter ξ_1 is determined from the relation $d = R$ ch ξ_1 or

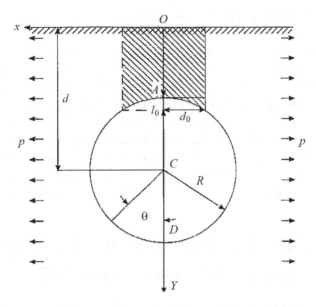

Fig. 4.22. Geometry of the problem of crack nucleation at a micropore with a radius R with the distance from the free face of the half plane being d, p is the external tensile pressure, $D = 2d_0$ is the crack thickness (crosshatched), θ is the angle counted from the positive direction of the Y axis in the clockwise direction; φ is the angle between the axis Y and the radius of the pore connecting its centre (point C) with the point of intersection of the upper (lower) surface of the crack and the pore (not shown in the figure).

$$\xi_1 = \ln\left\{\left[d + \left(d^2 - R^2\right)^{1/2}\right] / R\right\}. \qquad (4.41)$$

The relationship between the angle θ, counted in the clockwise direction from the positive direction of the Y axis, and the parameter η is determined by the equation

$$\sin\theta = x / R = \operatorname{sh}\xi_1 \sin\eta\left(\operatorname{ch}\xi_1 - \cos\eta\right). \qquad (4.42)$$

The abscissae of the points of the straight face of the half plane, coinciding with the X axis and perpendicular to the Y axis, are linked with the parameter η by the relation $x = a(1 - \cos\eta)/\sin\eta$, where $a = R\operatorname{sh}\xi_1$ or $a = d/\operatorname{cth}\xi_1$. The values of the coefficients P_n and N_n are presented in the form of tables for the actual values of ξ_1. The table gives the values of the stress concentration factors for $k_1 = \sigma_x/p$ and $k_2 = \sigma_\eta/p$ at the points O, A and D. Transferring from the mutually perpendicular, bipolar coordinate system ξ and η to the Cartesian system X and Y, to simplify the calculations, it is assumed that σ_x decreases in accordance with the linear law from the boundary of the half plane to the pore. The process of crack nucleation on the free surface to the pore is investigated taking into account the energy considerations (as in the case of the Griffith crack). The density of elastic energy is equal to $w_1 = (\sigma_x^2 + \sigma_y^2)/2E - v\sigma_x\sigma_y/E + (1 + v)\sigma_{xy}^2/E$. The formation of the crack is accompanied by the formation of two new surfaces and by disappearance of two surfaces whose elastic energy is equal to $w_2 = 294$ J·m^{-2}. The total energy of elastic deformation is equal to, as indicated by geometrical considerations

$$W_1 = 2\int w_1 ds \int (\sigma_x^2)/2E = \int [6.4y/(d-R)+0.314]^2 (p^2 d_0/E) dy +$$
$$+ p^2/E\left\{\left[d_0 l - \left[R(R-l)\sin 2\right]/2 + \pi R^2\varphi/360\right]\right\}, \qquad (4.43)$$

where the additional term, not included under the integral sign, is equal to, as indicated by the graph, to the difference between the area of the rectangle with the sides d_0 and l and the area of the segment. The surface energy is determined by the geometry of the problem and is equal to

$$W^2 = 294 \times 2 \times \left[(d-R+l)-d_0 - \pi R\varphi/180\right], \qquad (4.44)$$

and the thickness of the crack d_0 is determined from the condition of the minimum surface energy W_2; the upper and lower integration limits in equation (4.43) are equal to $(d - R)$ and 0, respectively.

The calculation show that the minimum of the surface energy is obtained at the crack thickness $D = 2d_0$ equal to $D \approx 1.16$ μm. Equating the relationships (4.43) and (4.44) gives the lower limit of the external stress p at which the formation of the crack connecting the pore with the free surface becomes efficient from the energy viewpoint, $p \approx 0.02$ GPa. Its value is 50–60 times lower than the yield stress (fracture stress) of the freshly quenched amorphous alloy ribbons. In conclusion, it should be mentioned that the value of p increases with increase of the distance of the pore from the free surface. This can be confirmed by carrying out quite simple but cumbersome calculations identical with those described above.

Analysis of the possibilities of crack nucleation in the ESL shows that cracks can also form at the inclusions with the linear size satisfying the ratio $R \geq l$, where R is any linear size of the inclusion which coincides with the elementary area normal to the external stress. The essential condition: the difference of the thermal coefficients of linear expansion of the amorphous inclusions and the ESL causing disruption of coherence at the interface between them. This case fully coincides with the nucleation of the crack at a pore.

Of special interest is the presence of an absolutely rigid inclusion situated in the amorphous matrix. If the relationship $R \geq l$ is fulfilled, the circumferential components of the stress tensor, like the stress concentration factor, tends to infinity at any external load p. This should result in fracture of the specimen if such inclusions form during the thermal–time influences.

The experiments also show that the initial stage of the ductile–brittle transition is characterised by the formation of ESL in capable of plastic deformation, with the physical and also mechanical characteristics of the material different from the properties of the amorphous centre. The critical size of the ESL will be estimated assuming that the length of the crack (the main mechanism of nucleation of the crack was discussed previously) is equal to the thickness of the ESL. According to [53], the general relationship for the critical rate of energy generation G_{1c} required for crack growth, has the form

$$G_{1c} = 2\,\gamma + 2l_{max}\,\eta_D, \qquad (4.45)$$

where γ is the specific surface energy of the propagating main crack; D is the energy scattered at the tip of the crack; l_{max} is the size of the zone of energy scattering normal to the crack surface. Equation (4.45) was obtained for the amorphous alloys deformed by the inhomogeneous mechanism with the formation of localised shear bands. G_{1c} is linked to the fracture toughness by the relationship $(K_{1c})2 = EG_{1c}$. The critical thickness of the ESL will be estimated using the results obtained in [54] where it is shown that in the bend test of the ribbon specimens of the amorphous alloy the maximum internal bending momentum M^{int} corresponds to the formation of a plastic hinge. In this case $M^{int} = (h\sigma_T d^2)/12$, where h and d are the width and thickness of the ribbon specimen, σ_T is the yield limit. Since the amorphous alloy is an elastoplastic medium with small strain hardening $m \leq 2$. The crack length will be assumed to be equal to the thickness of the ESL. According to [54], the crack of any length, adjacent to the plastic hinge, causes fracture of the specimen in the absence of strain hardening. Elemental calculations show that at the crack length of 2 μm (coinciding with the critical thickness of the ESL) the specimens fracture if $m \geq 0.5$ because the momentum of the internal stresses is not capable of compensating the external bending momentum. These considerations show that if the thickness of the ESL is smaller than the critical value, the cracks, nucleated in the ESL, close up after removing the load.

An intermediate case will now be investigated. In this case, the thickness of the ESL is greater than the critical value and the amorphous core is in the elastoplastic state. It has been shown that for the material in this condition the criteria of brittle fracture, obtained in the linear fracture theory, are not valid. In addition, in the temperature range of the ductile–brittle transition the type of deformation may change and this is associated with the transition from the inhomogeneous to homogeneous type. In the latter case when cooling the specimens of the amorphous alloy to room temperature the fracture must be brittle. The mechanical tests of the amorphous alloy to determine the microhardness and therefore the variation of the fracture toughness in the area of the ductile–brittle transition along the length of different cracks give values of K_{1c} averaged out over the thickness of the specimen. This is caused by the fact that to form the latter it is necessary to use maximum load during indentation, and the depth of the indentations is 7–9 μm. Thus, to determine the correlation between the fracture toughness K_{1c} and the relative strain ε_f it is necessary to carry out statistical processing of the measurement results. Investigation of different relationships between K_{1c} and ε_f shows that the lowest RMS deviation corresponded to the relation $\varepsilon_f \sim (K_{1c})^2$, and the correlation between these values was maximum. Thus, taking into account the dependence on the isochronous annealing temperature, the relationship between ε_f and K_{1c} can be written in the form

$$K_{1c}^2 \left(T_{a1} \right) / K_{1c}^2 \left(T_{a2} \right) = \varepsilon_f \left(T_{a1} \right) / \varepsilon_f \left(T_{a2} \right) \tag{4.46}$$

here T_{a1} and T_{a2} are the temperatures of isochronous annealing to which the specific values of K_{1c} and ε_f correspond. Therefore, the initial value $K_{1c} \left(T_{a1} \right) = 45$ MPa $\cdot$ m$^{1/2}$ corresponds to the relative fracture strain $\varepsilon_f \left(T_{a1} \right) = 1$ and then for $\varepsilon_f \left(T_{a2} \right)$:

$$\varepsilon_f \left(T_{a2} \right) = K_{1c}^2 \left(T_{a2} \right) / K_{1c}^2 \left(T_{a1} \right). \tag{4.47}$$

Substituting into equation (4.47) numerical values $K_{1c}^2 (T_{a1})=$ 45 MPa $\cdot$ m$^{1/2}$ and $K_{1c}^2 (T_{a2})=$ 10 MPa $\cdot$ m$^{1/2}$ gives $\varepsilon_f (T_{a2}) = 4.9 \cdot 10^2$. This value is almost identical with $\varepsilon_f(T_{a2})$, obtained by experiments for the $Co_{75.4}Fe_{3.5}Cr_{3.3}Si_{18.7}$ amorphous alloy.

In conclusion, attention will be given to the process of failure of the amorphous alloy after the ductile–brittle transition. This will be carried out using the accurate (taking into account the non-linear value of curvature) analytical solution of the problem of bending of the amorphous ribbon placed in the form of a half-loop between loading plates. It is assumed [55] that the contact or free surface of the ribbon is weakened by a slit (crack) with the length l. The stress intensity factors, produced by the slit, were obtained in [56–58]:

$$K_1 = 6M(d-l)^{-3/2} \zeta\big[(l/d)\big], \; K_{11} = 0, \qquad (4.48)$$

where M is the bending momentum at the tip of the bend, d is the thickness of the amorphous alloy ribbon, $\zeta[(l/d)]$ is the tabulated function with the numerical values of this function given in [58]. In [54] it was reported that the maximum value of the external bending momentum M in the purely elastic region per unit length of the amorphous ribbon, expressed through its Young modulus E and the geometrical characteristics, $M = 0.205E\varepsilon_f d^2$, where ε_f is the relative strain at fracture. Substituting the value of M into equation (4.48) and after relatively simple transformations, the resultant expression for the critical crack length l_{cr} is:

$$l_{cr} \leq d - \Big[1.23 E \varepsilon_f d^2 \zeta\big[(l/d)\big]/K_{1c}\Big]^{2/3}. \qquad (4.49)$$

Substituting into the inequality (4.49) the numerical values of the elastic and geometrical characteristics of the ribbon of the $Co_{75.4}Fe_{3.5}Cr_{3.3}Si_{18.7}$ amorphous alloy and also the experimental values of K_{1c} and ε_f, after completing the ductile–brittle transition the critical crack length is $l_{cr} = 1.5$–2 µm. Similar values were also obtained for the amorphous alloy with the composition $Fe_{61}Co_{20}B_{14}Si_5$.

Thus, the results obtained in this section may be summarized as follows:

- the critical size of the ESL;
- the functional dependence of the relative fracture strain in bend tests of the ribbon specimens of the amorphous alloy in relation to the fracture toughness which, in turn, depends on the

isochronous annealing parameter;
• confirmation that the material of the ESL is in the amorphous
  state which differs from the initial state and is characterised in
  particular by the minimum fracture toughness K_{1c};
• the value of the critical crack length of the material of the
  amorphous alloy which after the ductile–brittle transition
  coincides with a high degree of accuracy with the critical
  thickness of the ESL;
• the data confirming that when the intensity of the thermal–time
  effects increases, the material of the ESL overlaps the entire
  cross-section of the amorphous ribbon.

The theoretical and experimental investigations of the phenomenon
of the ductile–brittle transition in the amorphous alloys of the metal–
metalloid system indicate that:
  1. Structural relaxation is one of the most important reasons
for the ductile–brittle transition. This transition can not take place
without segregation processes, including the transfer of the excess
part of the free volume through the surfaces of the amorphous
ribbons and displacement into the micropores. The atoms of the
metalloids are also transferred to the subsurface regions of the ribbon
forming segregates at the micropores in the volume of the amorphous
alloy and in the vicinity of the surface.
  2. The ductile–brittle transition is a process which takes place
in time. Investigation of the kinetics of this process has shown that
the formation of the ESL which have lost their capacity for plastic
shape changes is the initial stage of the process. The increase of
the intensity of the thermal–time effects increases the thickness of
the ESL up to the moment when the latter do not overlap the entire
cross-section of the ribbon.
  3. The most important macroscopic manifestation of the ductile–
brittle transition is the catastrophic decrease of the fracture stress by
a factor of 20–25 and of the fracture toughness by 3.5–4 times. At
the standard duration of isochronous annealing of $\tau = 10$ min, the
transfer of the amorphous alloy material from the ductile to brittle
state takes place in a narrow temperature range of 30 to 50 K and
depends on the composition of the amorphous alloy.
  4. The classic (linear), momentum and micropolar elasticity theory
and linear and non-linear fracture mechanics were used to determine
the critical thickness of the ESL equal to ≈ 2 µm and also determine

the general relationships linking the decrease of the fracture strain with a decrease of the fracture toughness of the amorphous alloy.

5. The mechanism of crack nucleation and micropores adjacent to the ESL has been proposed and the conditions of propagation and transformation to main cracks after thermal and destabilizing effects have been outlined. The quantitative and qualitative agreement between the acoustic emission events, the characteristics of the individual pulses and their statistics, obtained in deformation of the freshly quenched ribbon specimens of the amorphous alloy (Poisson statistics) and embrittled specimen (of the correlation nature, typical of brittle fracture) has been determined.

6. A new characteristic material constant of the amorphous alloy was determined. This constant forms in the momentum theory of elasticity and comparison with this constant of the linear dimensions of inhomogeneities, including the micropores, can be used to separate these defects and also the stress concentration factors characterising the fields of additional elastic stresses generated by them during acting loading, to macroscopic and local.

7. The model of the amorphous alloys as a multilevel system with the continuous distribution of the spectrum of the activation energies has been constructed and analytical expressions have been derived for the shape of the heat generation spectrum in the scanning and isothermal conditions. It was therefore possible to formulate the criterion of the ductile–brittle transition of the amorphous alloy in the close correlation between the evolution of the relaxation spectrum and the variation of the mechanical behaviour of the amorphous alloy determined for the first time.

8. The proposed physical model of the ductile–brittle transition can be used to explain the effect of the conditions of surface-active elements and the atoms of rare-earth elements on the displacement of its temperature and time boundaries. Since the surface active elements decrease the surface energy, this facilitates the nucleation and propagation of main microcracks. The rare-earth elements bond the free atoms of the metalloids preventing the formation of segregations of the latter in the subsurface layers and in the volume of the amorphous alloy and decrease at the same time the temperature of the ductile–brittle transition to the range of pre-crystallisation temperatures.

References

1.   Glezer A.M. et al., Structural causes of temper embrittlement of amorphous alloys of metal - metalloid, FMM. - 1984. - V. 58. - Vol. 5. - P. 991-1000.
2.   Utevsky L.M., et al. Reversible temper brittleness of steel and iron alloys. - Moscow: Metallurgiya, 1987.
3.   Chen H.S. Ductile-brittle transition in metallic glasses, Mater. Sci. and Eng. - 1976. - V. 26. - No. 1. - P. 79-82.
4.   Naka M., Masumoto T., Chen H.S. Effect of metalloidal elements on strength and thermal stability of iron-base glasses, J. Phys. - 1980. - V. 41. - No. 8. - P. C8-839-C8-842.
5.   Stubicar M. Microhardness characterization of stability of Fe-Ni-base metallic glasses, J. Mater. Sci. - 1979. - V. 14. - No. 6. - P. 1245-1248.
6.   Bresson I., Harmelin M., Bigot J. Influence of phosphorus and sulphur additions on the mechanical, surface and thermal properties of $Fe_{78}B_{13}Si_9$ amorphous alloy. J. Mat. Sci. and Eng. - 1988. - V. 98. - No. 2. - P. 495-500.
7.   Egami T. Structural relaxation in amorphous alloys and compositional short range ordering, Mater. Res. Bull. - 1978. - V. 13. - No. 6. - P. 557-562.
8.   Deng D., Argon A.S. Effect of aging on distributed shear relaxation, hardness and embrittlement in $Cu_{59}Zr_{41}$ and embrittlement in $Cu_{59}Zr_{41}$ and $Fe_{80}B_{20}$ glasses, Proc. Fifth Int. Conf. RQM. Elsevier Sci. Publ. - 1985. - V. 2. - P. 771-774.
9.   Yamasaki L., Takahoshi M., Ogino Y. Compositional dependence of temper embrittlement of some Fe-based amorphous alloys, Proc. Fifth Int. Conf. RQM. Elsevier Sci. Publ. - 1985. - V. 2. - P. 1381-1384.
10.   Chen H.S. Thermal and mechanical stability of metallic glass ferromagnets, Scr. Met. - 1977. - V. 11. - No. 5. - P. 367-370.
11.   Misra R.D.K., Akhtar D. Annealing effects in $Ni_{60}Nb_{40-x}Al_x$ metallic glasses, Mater. Sci. and Eng. - 1987. - V. 92. - No. 2. - P. 207-216.
12.   Latuszkiewicz J., Zielinski F.G., Matyja H. Ductile-to-brittle transition in Fe-Ni-Si-B metallic glasses, Proc. Int. Conf. Metal. Glass .: Science and Technology. (Budapest, Hungary). - 1980. - P. 283-289.
13.   Chen H.S. Correlation between structural relaxation processes and the kinetics of thermal and mechanical stabilities in metallic glasses, Proc. Fourth Int. Conf. RQM. (Sendai, Japan). - 1981. - V. 1. - P. 555-558.
14.   Gerling R., Schimznsky F.P., Wagner R. Influence of the thickness of amorphous-$Fe_{40}Ni_{40}B_{20}$ ribbons on their mechanical properties under neutron-irradiation and thermal annealing, Proc. Fifth Int. Conf. RQM. Elsevier Sci. Publ. - 1985. - V. 2. - P. 1377-1380.
15.   Zaichenko S.G., Braginskii A.P. Metallofizika. - 1990. - V. 12. - No. 4. - P. 15-21.
16.   Zaichenko S.G., Aldokhin DV, Glezer A.M. Physical model of ductile-brittle transition in amorphous metal-metalloid alloys: thermodynamic description and acoustic emission research, Izv. RAN Ser. Fiz. 2005. - V. 69. - No. P. - S. 1363-1368.
17.   Neuhäuser H. Rate of shear band formation in metallic glasses, Scr. Met. - 1978. - V. 12. - No. 5. - P. 471-474.
18.   De Zhen J. Physics of Liquid Crystals. - Moscow: Mir, 1977. - 400 p.
19.   Glezer A.M. Betekhtin V.I. The free volume and mechanisms of microfracture of amorphous alloys, Fiz. Tverd. Tela - 1996. - V. 38. - No. 6. - S. 1784-1790.
20.   Komatsu T., Matusida K., Yokota R. Structural relaxation and embrittlement in Fe-Ni based-metallic glasses, J. Mater. Sci. - 1985. - V. 20. - No. 8. - P. 1376-1382.
21.   Pampillo C.A., Polk D.E. Annealing embrittlement in an iron-nickel-based metallic

glasses, J. Mater. Sci. Eng. - 1978. - V. 33. - No. 2. - P. 275-280.

22. Fujita F.E. On the intermediate range ordering in amorphous structure, Proc.Fourth Int. Conf. RQM. (Sendai, Japan). - 1981. - V. 1. - P. 301-304.

23. Egami T. Structural relaxation and magnetism in amorphous alloys., J. Magn. and Magn. Mater. - 1983. - V. 31-34. - Part 3. - P. 1571-1574.

24. Inoue A., Kobayashi K., Nose M., Masumoto T. Mechanical properties of (Fe, Co,Ni) -M-B (M-Ti, Zr, Hf, V, Nb, Ta and Mo) amorphous alloys with low boron-concentration, J. Phys. - 1980. - V. 41. - No. 8. - P. C8-831-C8-834.

25. Kimura H., Masumoto T. Deformation and fracture of an amorphous Pd-Cu-Si alloy in V-notch bending test. I. Model mechanics of inhomogeneous plastic flow in non-strain hardening solids, Acta Metal. - 1980. - V. 28. - No. 7. - P. 1663-1675.

26. Kimura H., Masumoto T. Deformation and fracture of an amorphous Pd-Cu-Si alloy in V-notch bending test. - II. Ductile-brittle transition, Acta Met. - 1980. - V. 28. - No. 7. - P. 1677-1693.

27. Kimura H., Masumoto T. Fracture mechanics for embrittlement of a Fe-base amorphous alloy., Proc. Fifth Int. Conf. RQM. Elsevier Sci. Publ. - 1985. - V. 2. - P. 1373-1376.

28. Liebermann H.H., Luborsky F.E. Embrittlement of some metallic glasses by Sb, Se and Te, Acta Met. - 1981. - V. 29. - No. 6. - P. 1413-1418.

29. Spaepen F., Turnbull D. A mechanism for the flow and fracture of metallic glasses, Scr. Met. - 1974. - V. 8. - No. 2. - P. 563-568.

30. Wu T.W., Spaepen F. Small angle X-ray scattering from an embrittling metallic glass, Acta Met. - 1985. - V. 33. - No. 11. - P. 2185-2187.

31. Spaepen F., Taub A.I., Plastic flow and fracture, Amorphous metal alloys. - Moscow: Metallurgiya, 1987. - P. 228-256.

32. Yavari A.R. Formation of boron-rich zones and embrittlement of Fe-B-type metallic glasses, J. Mater. Res. - 1986. - V. 1. - No. 6. - P. 746-751.

33. Pratten N.A., Scott M.G. Stability of some metalloid-free metallic glasses, Scr. Met. - 1978. - V. 12. - No. 2. - P. 137-142.

34. Sonius M.E., Thijsse B.Y., Benkel A. The kinetics of structural relaxation in amorphous $Fe_{40}Ni_{40}P_{14}B_6$, Scr. Met. - 1983. - V. 17. - No. 4. - P. 545-548.

35. Yagielinski T., Egami T. Reversibility of the structural relaxation in amorphous alloys, J. Appl. Phys. - 1984. - V. 55. - No. 6. - P. 1811-1813.

36. Muskhelishvili NI Some basic problems of the mathematical theory elasticity. - Moscow: Publishing House of the USSR Academy of Sciences, 1954. - 647 p.

37. Onami M. et al. Introduction into micromechanics. - Moscow: Metallurgiya, 1987.

38. Timoshenko S.P. , Theory of elasticity. - Moscow: Nauka, 1979.

39. Neuber N., Ingenieur - Arch. - 1934. - V. 5. - P. 242.

40. Zaichenko S.G., Glezer A.M. Physical model of ductile-brittle transition amorphous alloys of the metal–metalloid type: heterogeneity of structure, creating a field of elastic stresses in the active loading, Izv. RAN. Ser. Fiz. - 2004. - V. 68. - No. 10. - P. 1488-1494.

41. Savin GN Stress distribution near holes. - Kiev - Naukova Dumka, 1968.

42. Braginsky A.P. et al. Pis'ma ZhTF. - 1986. - V. 12. - No. 18. - P. 1111-1114.

43. Zaichenko S.G., Braginsky A.P., Disclinations and rotational ductility of solids. - Lenigrad: A.F. Ioffe Institute. 1990. - S. 140.

44. Zaichenko S.G., Braginsky A.P., Metallofizika. - 1990. - V. 12. - No. 4. - P. 15.

45. Braginsky A.P., Physics and Mechanics of Composite Materials. -Lenigrad: A.F. Ioffe Institute. 1986. - P. 10.

46. Glezer A.M., Zaichenko S.G. Izv. RAN. Ser. Fiz. - 2006. - V. 70. - No. 9. -P. 1366-

1371.
47.   Zaichenko S.G., Kachalov V.M., Mater. Sci. Eng. A. - 1977. - V. 226-228. - P. 353.
48.   Zaichenko S.G, Glezer A.M. Izn RAN. Ser. Fiz. - 2006. - V. 70. - No. 9. - S. 1359-
       1365.
49.   Zaichenko S.G., Perov N.S., Ganshina E.A., J. Phys. (France). - 1998. - V. 8. - P. 2.
50.   Poperenko L.V., Vinnichenko M.V., Zaichenko S.G., et al., Proc. M.R.S. Symp. -
       1996. - V. 400. - P. 335.
51.   Poperenko L.V., et al., Avtometriya. - 1997. - No. 5. - P. 76.
52.   Jeffry G.B., Trans. Roy. Soc. (London). A. - 1921. - V. 221. - P. 265.
53.   Krener W., Pompe W., J. Mater. Sci. - 1981. - V. 16. - P. 694.
54.   Zaichenko S.G., et al. Zavod. Lab. - 1989. - No. 5. - P. 76.
55.   Broek, D. Fundamentals of fracture mechanics. - Moscow: Vysshaya shkola, 1980.
56.   Bueckner H.F., Boundary problems in differential equations. - Wisconsin: Univ.
       Wisconsin Press, 1960. - P. 215.
57.   Paris P., Sih G. Applied problems of fracturetoughness / ed. M.A. Shtremel'. - Mos-
       cow: Mir, 1968. - P. 174.
58.   Cherepanov G.P. The mechanics of brittle fracture. - Moscow: Nauka, 1974.

5

Physical criteria for predicting thermal and time stability

Previously, the thermal and time stability of amorphous alloys was predicted using the diagrams of thermal and time stability (DTTS) [1] (Fig. 5.1). To construct the DTTS, the amorphous alloys were subjected to isothermal annealing at different thermal and time parameters. The ductile–brittle transition was determined using the parameter ε_f in mechanical tests at room temperature with 180° free bending [2]. The lower boundary of the ductile–brittle transition was determined by brittle fracture of at least one specimen from a set with the total number of the specimens processed in the given

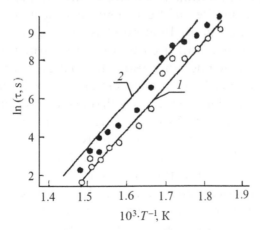

Fig. 5.1. Diagram of the thermal and time stability for the Fe–Co–Si–B amorphous alloys; the numbers indicate the lines approximating the lower (curve 1) and upper (curve 2) boundaries of the ductile–brittle transition.

conditions of $N = 10-15$; the upper boundary was determined by brittle fracture of all specimens of the set. Both the upper and lower boundaries of the ductile–brittle transition depend on the specific composition of the amorphous alloy and are described quite satisfactorily by the Arrhenius dependence of the type

$$\tau_{br} = \tau_0 \, \exp(U/RT), \tag{5.1}$$

where τ_{br} is the time to the ductile–brittle transition after annealing at temperature T, K; U is the activation energy of the process of the ductile–brittle transition which can be determined by the Kissinger [3] or Ozawa [4] method from the temperature displacement of the heat generation peak of the relaxation spectrum with the variation of the heating rate of the specimen; τ_0 is the pre-exponential multiplier.

Unfortunately, the methods of constructing the DTTS can be used only for determining the thermal–time parameters of the ductile–brittle transition in the range of high temperatures (close to the glass transition temperature T_g) and relatively short time periods. Consequently, it cannot be used for predicting the possibility of the ductile–brittle transition in the climatic (less than 373 K) temperature range at very long (possibly years) times. This is associated with the following main reasons [5]:

1. The approximating dependence (5.1) does not have the strict theoretical and experimental justification.

2. Similar approximation is possible only if there is a single mechanism of the ductile–brittle transition in a wide temperature and time range which in all likelihood is not realistic because the transition from the region of high temperature structural relaxation to the region of low-temperature relaxation takes place.

3. If the two previously mentioned circumstances do not introduce any significant restrictions (which is unlikely), then even in this case in a number of amorphous alloys there is no distinctive peak of the relaxation spectrum and this results in unavoidable serious errors in the expression for the exponent in equation (5.1).

4. Determination of the thermal and time parameters of the ductile–brittle transition purely empirically is not realistic because it requires thermal effects at the given temperature in the climatic range over not hundreds but thousands of hours.

Therefore, the following tasks should be fulfilled:

a) the development of scientifically justified methods of predicting the time stability of the mechanical properties of amorphous alloys

Table 5.1. Chemical composition and main characteristics of the investigated alloys

Notation	Composition, at.%	T_s, K	T_{br}^0, K	k, deg/h
Alloy 1	$Fe_{70}Cr_{15}B_5$	788	593	12.0
Alloy 2	$Fe_{70}Cr_{15}P_7B_8$	808	493	7.7
Alloy 3	$Fe_{76}Ni_2Si_9B_{13}$	818	473	7.0
Alloy 4	$Fe_{76}Ni_2Si_9B_{13}$	823	553	10.5
Alloy 5	$Fe_{46}Co_{32}Si_9B_{13}$	793	573	10.0

(the moment of the start of the ductile–brittle transition) in the climatic temperature range;

b) determination of the method of the possible increase of the thermal and time stability (displacement of the moment of the start of the ductile–brittle transition to higher temperatures and longer annealing times).

To develop the physically substantiated method of predicting the ductile–brittle transition, the main criterion of the thermal and time stability of the amorphous alloys in [6, 7] was developed by investigations on five amorphous alloys of the transitional metal–metalloid type, produced by the spinning method (Table 5.1).

T_{br} – the criterion of the ductile–brittle transition. As mentioned previously, the main parameter characterising the ductile–brittle behaviour of amorphous alloys under the thermal effects is the temperature of the ductile–brittle transition (temper brittleness) T_{br}. Therefore, it is logical to attempt to use this characteristic for predicting the ductile–brittle transition in low-temperature relaxation in the climatic temperature range under long-term thermal effects.

The 'reference' value of T_{br}, with the value of this parameter being the starting point for determining the thermal and time stability of the amorphous alloys, is represented by the temper brittleness temperature after isochronous annealing for one hour. This characteristic which will be denoted by T_{br} is a physical constant for the amorphous alloy of the given composition and is also listed in Table 5.1. If the annealing (storage) temperature T_0 is lower than T_{br}^0, the ductile–brittle transition in the amorphous alloy takes place during the time τ longer than 1 hour. In a general case, when $T_0 < T_{br}^0$, the susceptibility to the ductile–brittle transition will constantly change (increase) with increase of the annealing time τ. The value of T_{br} will also change (decrease) in a corresponding manner at every moment of time. Finally, a moment may arise when $T_{br} = T_0$ and the alloy undergoes the ductile–brittle transition. Thus, the value T_{br} for the given alloy in a specific structural state is a

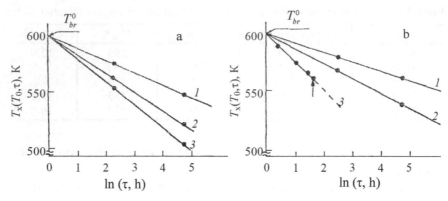

Fig. 5.2. Dependence of T_{br} on the duration of preliminary isothermal annealing τ_0 for the alloy 1 (a) and alloy 2 (b) at temperatures of 333 K (curve 1), 393 K (curve 2) and 453 K (curve 3).

measure of its susceptibility to transition to the brittle state: as the value of T_{br} becomes closer to T_0 the alloy will be in the ductile state for shorter and shorter periods of time.

Figure 5.2 shows the resultant dependences $T_{br} = f(\ln \tau)$ for the alloys 1 and 2 at different annealing temperatures. Similar dependences also form for the other alloys investigated in [7]. It may be seen that the dependence of the type

$$T_{br} (T_0, \tau) = C \ln \tau \qquad (5.2)$$

which is well-known from the literature [8] for high-temperature relaxation ($T0 \approx T_g$), is satisfied quite well also in the case of low-temperature relaxation ($T_0 \ll T_g$). It should only be assumed that the dependence holds in the entire range of the values $T_0 \ll T_g$. The straight-line in Fig. 5.2b breaks when T_{br} and T_0 become identical, i.e., when the ductile–brittle transition starts to take place.

The claim according to which the dependence (5.2) applies in the entire temperature range $T_0 < T_{br}^0$ is based on the fact that this temperature range is in the range of low-temperature relaxation which can be described by the same relationships on the condition that T_0 is considerably lower than T_g.

The problem is reduced to the need to determine correctly the type of approximating function $C(T)$. Computer processing of the dependences of the type $T_{br} = C \ln \tau$, obtained at different temperatures, shows that for all the investigated alloys the approximating function $C(T)$ has the form

$$C(T) = k\eta(1-\eta)^{-0.5}, \tag{5.3}$$

where $T0/T_{br}^0$, $k = $ const.

Equation (5.3) can now be written in the form

$$T_{br}^0 - T_{br}(T_0, \tau) = k\left(\eta(1-\eta)^{-0.5}\right)\ln\tau, \tag{5.4}$$

where k has the meaning of the kinetic factor of the ductile–brittle transition. After introducing the notation $\Delta T_{br}^0 = T_{br}^0 - T_{br}(T_0, \tau)$ the value τ is expressed from the equation (5.4)

$$\tau_0 = \exp\left[\Delta T_{br}(1-\eta)^{-0.5} / (k\eta)\right]. \tag{5.5}$$

The equations (5.4) and 5.5) make it possible, on the one hand, to estimate the decrease of the temper brittleness temperature during annealing at T_0 (in particular, at room temperature) and, on the other hand, the time required for this decrease. Independent experiments must be carried out to determine the value of the kinetic factor k which is characteristic of every specific amorphous alloy. Its value can be easily determined from the graphs shown in Fig. 5.3 which show the dependence of the coefficient $C = \Delta T_{br}/\ln\tau$ on the parameter $\eta(1 - \eta)^{-0.5}$. It may be seen that each of the investigated alloys has its own value of k, and the numerical values of this

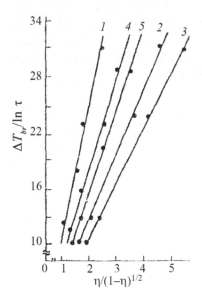

Fig. 5.3. Dependence of $\Delta T_{br}/\ln\tau$ on $\eta(1 - \eta)^{-0.5}$ for the investigated alloys.

Table 5.2. The time to the start of the ductile–brittle transition (τ_0, years) of the investigated amorphous alloys at room temperature, calculated using the independent T_{br} and RS-criteria

Criterion	Amorphous alloy				
	1	2	3	4	5
T_{br}-criterion	25.4	8.0	6.2	9.1	45.5
RS-criterion	26.7	8.9	7.0	8.7	50.0

parameter are presented in Table 5.1. Analysing the dependences presented in Fig. 5.3 it can be shown that the values of k for the alloys containing phosphorus are slightly lower than those obtained for the alloys containing boron only. In practice, the value of k can be determined by means of simple experiments: record the initial value of T_{br}^0, typical of the given alloy, and then the value T_{br} after annealing at the given temperature T_0 for time τ. Evidently, the accuracy of determination of k will increase with the increase of the values of $\Delta T_{br} = T_{br}^0 - T_{br}^\tau$ and τ.

Taking these considerations into account one can predict the time period during which the investigated alloys will not undergo the ductile–brittle transition at room temperature. The results of calculations based on the T_{br}-criterion of the ductile–brittle transition, are presented in Table 5.2. In may be seen that the time stability of the ductility is maximum for the $Fe_{70}Cr_{15}B_{15}$ and $Fe_{46}Co_{32}Si_9B_{13}$ alloys. It should be noted that this prediction was made assuming that the alloys would not change their chemical and phase composition during isothermal annealing. This assumption is fully justified and was confirmed by controlling the chemical composition of the surface of ribbon specimens by Auger spectroscopy in different stages of the thermal and time effects in the climatic temperature range [6, 7].

The RS-criterion of the ductile–brittle transition. One of the efficient methods of obtaining information about the physical and structural parameters of the amorphous alloys is the method of differential scanning calorimetry (DSC) and differential thermal analysis (DTA). These methods can be used to investigate the evolution of heat generation of the relaxation spectrum (RS) in dependence on the intensity and nature of the destabilising effects, in particular after the thermal and time effects on the amorphous alloys [6]. The thermodynamic approach defines a general direction of the evolution of the amorphous alloys as a system that is in the metastable equilibrium state, without specifying the macroscopic mechanisms responsible for this process. The concentration factors of elastic stresses, produced in [9], with the elastic stresses created by the heterogeneities of the structure of the amorphous alloys in active loading, will be used in analysis of the specific mechanisms of structural relaxation and plastic deformation and the processes of micro- and macroscopic fracture.

The integral characteristic of the deviation of the amorphous alloy from the thermodynamic equilibrium state is the heat generation of the relaxation spectrum (S). The process of evolution of the RS will be investigated during isochronous annealing with the increase of annealing temperature and will be compared with the variation of the mechanical properties of the amorphous alloy. To simplify the considerations, investigations will start with the analytical description of the RS of the two-level system.

It is assumed that at the atomic level the process of structural relaxation (according to the classic scheme in [10]) is based on the redistribution of the atoms between two energy levels with the energies U_1 and U_2 ($U_1 > U_2$) as a result of the passage of the particles through the potential barrier separating them. The population of the upper level ξ of the two-level system is described by the kinetic equation (disregarding the entropy terms) [10]

$$d\xi/dT = \omega(\eta - \xi), \tag{5.6}$$

where $\omega = \omega_{12} + \omega_{21}$ is the total frequency of transitions of the particles from the upper to lower level and back. According to the theory of absolute reaction rates, the probabilities of the transition of the atoms from the upper to lower level ω_{12} and from the lower to upper level ω_{21} have the form

$$\omega = v_1 \exp(-U_1/RT) + v_2 \exp(-U_2/RT), \tag{5.7}$$

where v_1 and v_2 are the frequencies of transitions of the atoms between the levels 1 and 2; 2 and 1, respectively. Equation (5.6) shows that the process is determined by the specific relaxation rate $\omega \leq \omega_{12} + \omega_{21}$ (ω^{-1} is the relaxation time) and by the deviation of the distribution ξ from the thermodynamically equilibrium distribution characterised by the term $\eta = \omega_{12}/\omega$. The exact solution of the relaxation equation has the form

$$\xi(z) = \xi_0 \exp(-\tau) + e \cdot \exp(-\tau) \times \int_0^z \eta w \exp\left[\tau(z')(dz')\right], \qquad (5.8)$$

where $\tau = \int_0^z w \, dz'; \; w = \omega / q; \; q = dz / dT; z = kT / U_2;$

z and q are the dimensionless temperature and heating rate (cooling rate). The temperature dependence of heat generation (disregarding the entropy terms) has the form

$$i = -\Delta U \, d\xi/dT \qquad (5.9)$$

and in the scanning mode (dT/dt = const) is described by the equation

$$di / dt = -\Delta U d \left(\omega [\eta - \xi]\right) / dt = i \cdot \omega \left(y^{-1} - 1\right), \qquad (5.10)$$

where $y = \omega RT^2/(U\dot{T})$. When deriving the equation (5.10) it is assumed that $\Delta U = U_1 - U_2$ is a small value, i.e., $U_1 \approx U_2 = U$. In addition, since the parameter η is a slowly changing function of time ($\eta \sim 10^{-2}$) with the accuracy sufficient for practical calculations, as indicated by the experimental data, it can be ignored. The solution of equation (5.10) with this accuracy has the form

$$i(T) = i_{max} \, y \, \exp \, [1 - y], \qquad (5.11)$$

because for any solution of the equation (5.10) the maximum value is obtained at the point $y = 1$.

Further, it is assumed that the spectral density of heat generation RS $h(T, T_a, T)$, where T is actual temperature, T_a is the annealing temperature, t is time, is an additive sum of the elementary two-level transitions. If $F(U)$ is the distribution function of the activation energies U, then for $h(T, T_a, t)$

$$h(T,T_a,t) = \int_U I_{max} y(z) \left[\exp(1-y)\right] dU, \qquad (5.2)$$

here $I_{max} = F(U)/e$ has the meaning of the maximum total heat generation produced by all two-level systems with the activation energy U, $\omega \approx \omega_{12} + \omega_{21} \approx 2$ is the total frequency of the transitions through the energy barrier U and in the reverse direction. Since the function $y \times \exp(1 - y)$ in the scanning mode has a very sharp maximum, the main part of the integral (592) in this case can be presented in the form [11]

$$\int_T h(T,T_a,t) = 2^{1/2} RTU_{max},$$ (5.13)

and in the isothermal mode

$$\int_T h(T,T_a,t)dT = \left(\int_{T_0}^{T^*}(T,T_a^*,t)dT\right)/\left[1+x^*\dot{T}(t-t^*)/T^*\right],$$ (5.14)

where the upper index (*) denotes the transition from the scanning to isothermal mode.

There are various methods for determining the activation energies of heat generation of the RS. One of them [3] is based on the displacement of the heat generation peak of the RS $H(T_a,t) = \int_T h(T,T_a,t)dT$ with the variation of the heating rate. This method may be described as follows. The main part (5.13) separated from the integral (5.12) links the heat generation $H(T_a, t)$ of the RS, the temperature T and maximum heat generation $I_{max}(U)$ with the activation energy U:

$$U = x^*RT.$$ (5.15)

Parameter x^* is determined from the equation

$$x^*\exp(x^*) = 2vT/\dot{T}.$$ (5.16)

The last equation follows from the condition of the maximum value of the heat generation of the two-level system with the energy U^*. Further, from the Kissinger equation [3]

$$d(\ln T_p)/d(\ln \dot{T}_p) = U^*/RT_p$$ (5.17)

it is possible to determine the activation energy at the peak of the complex relaxation spectrum as a result of the variation of $\dot{T}$. The combined solution of the equations (5.15)–(5.17) gives the values of the activation energies adjacent to the peak with the subsequent construction of the energy spectrum I_{max} (U) for any temperature T.

In contrast to the condensed systems, characterised by the translational symmetry, all the atoms of the amorphous alloy are in the excited state. This is confirmed by the experiments with the heat generation of the relaxation spectra in the temperature range from 300 K to the pre-crystallisation temperatures in heating with any rate (Fig. 5.4). The relaxation spectra of the amorphous alloy of the M–M type have the following special features:

1. Gradual disappearance with increasing annealing time of the low-temperature (in the temperature range from $T \approx 393$ to ≈ 668–673 K) part of the RS, whereas the high-temperature part (situated in the range $(T \approx 668$–673 K$) - T_g$, where T_g is the glass transition temperature) remains almost unchanged (Fig. 5.4);

2. The variation of the total heat generation of the low temperature part of the relaxation spectra (LTRS) after annealing, including storage at room temperature, normalised for the heat generation of the LTRS of the freshly quenched material of the amorphous alloy, as indicated by the equation (5.16) is proportional to

$$\int_T h(T,T_a,t)dT \, / \int_T h(T,0,0)dT = 1 - k\cdot\ln t, \qquad (5.18)$$

where k is a function of T_a, x^*, U, $\dot{T}$;

3. Annealing is accompanied by the change of the sign in the individual sections of heat generation of the LTRS, and the amorphous alloy absorbs heat (Fig. 5.4, curve 3). Comparison of the mechanical test results of the ribbon specimens of the amorphous

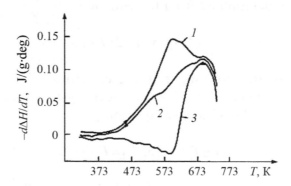

Fig. 5.4. Typical relaxation spectra of the amorphous alloy of the metal–metalloid type (Fe–Cr–B system (b)), obtained by the DSC method at a heating rate of dT/dt = 0.3 K/s. Notations: curve 1 – RS of the freshly quenched ribbons; curve 2 – the RS after storage at room temperature T = 293 K for 8000 h, curve 3, – RS after low-temperature annealing with the parameters: T = 493 K, t_0 = 100 h.

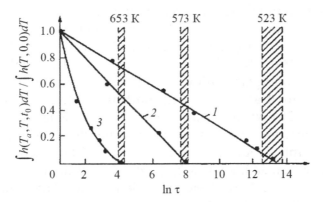

Fig. 5.5. Dependence of the relative heat generation of the amorphous alloy (a) on the duration τ of isochronous annealing at temperatures: $T_a = 523$ K (curve 1), $T_a = 573$ K (curve 2), and $T_a = 653$ K (curve 3). The crosshatched areas between the vertical lines indicate the lower and upper boundaries of the ductile–brittle transition.

alloys subjected to annealing at different temperatures and duration with the evolution of the heat generation in the LTRS of the same specimens can be used to formulate the criterion of the ductile–brittle transition: the amorphous alloy is transferred from the ductile to embrittled state under the condition of equality to 0 of the integral regeneration of the LTRS, including sections with heat absorption:

$$\int_T h(T_a, T, t) dT = 0. \tag{5.19}$$

It should be noted that equation (5.14) shows that to predict the thermal stability of the mechanical properties of the amorphous alloys of the M–M type (at a fixed annealing temperature) it is sufficient to compare the heat generation of two RSs produced by differential scanning calorimetry (DSC). One is the material of the freshly quenched ribbons of the amorphous alloy $H(T, 0)$, the other one after appearance of the visible changes of the RS during annealing with the duration t_0 $H(T_a, t_0)$. The time to the start of the ductile–brittle transition is determined by the equation

$$t_{br} = t_0 \exp\left[\left[1 - \int_t h(T_a, T, t_0) dT \Big/ \int_t h(T_a, 0, 0) dT\right]^{-1} - 1\right]. \tag{5.20}$$

The embrittlement criterion (5.19) together with the equation (5.20) which can be used to predict the time to the start of the ductile–

brittle transition taking into account the thermal prior history of the amorphous alloy, was verified by experiments on amorphous alloys of the Fe–Co–Si–B (a), Fe–Cr–B (b), Fe–Cr–B–P (c), Fe–Cr–B–Al (d), Fe–Ni–B–Si (e), Fe–Co–Ni–Si–B (f) and a number of other alloys (~25 amorphous alloys of different composition) [6]. Figure 5.6 shows good agreement with both the criterion (5.19) and the equation (5.20) which makes it possible to predict the start of the ductile–brittle transition taking into account the thermal prior history of the amorphous alloys. However, it should be noted that in high-temperature annealing when $T_a \geq 647$ K, and in short-term holding the curve of the relative heat generation of the LTRS demonstrates a more complicated dependence on ln (t_0). This can be obtained from the equation (5.18) taking into account all the terms after integration which are not important in annealing of medium and long duration.

The specimens prepared from the amorphous alloy ribbons of the systems (a–e) were processed by low-temperature annealing (artificial ageing at $T = 453$ K). The heat generation of RS was investigated by the DSC method in a Dupon thermal analyser with the scanning speed of $T = 0.33$ K/s. Figure 5.4 shows the dependences of the heat generation $H(T_a, t)$ of the RS of the alloy of the Fe–Cr–B system in the freshly quenched state (curve 1) after storage at room temperature (curve 2) and low-temperature annealing (artificial ageing) (curve 3). The nature of heat generation of the RS $H(T_a, d)$ of the remaining alloys is similar to that shown in Fig. 5.4. Curve 2 in Fig. 5.4 shows that storage had a comparatively weak effect on the form of $h(T_a = 293$ K, $T, t)$ in comparison with the radical changes of the spectral density of heat generation after low-temperature annealing (curve 3). These changes were reflected in the complete disappearance of the heat generation of the LTRS and in the appearance of areas with absorption of heat up to the temperatures of $T = 648$–653 K. The storage and low-temperature annealing had no effect on the quantitative and qualitative characteristics of the spectral density of heat generation $h(T_a, T, t)$ of the high-temperature part of the RS. The mechanical tests of the ribbon specimens by 180° free bending [2] were carried out to check the state of the amorphous alloy after storing. None of the amorphous alloys of the systems (a–e) showed any loss of ductility, but after annealing the amorphous alloys of the systems c and d showed the ductile–brittle transition; the amorphous alloy of the system e was in the semi-brittle state; b and f demonstrated ductility, as in the freshly quenched ribbons. It should be noted that the term 'semi-embrittlement' is used in

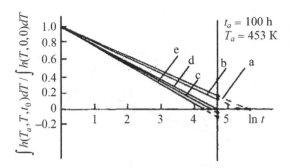

Fig. 5.6. Correlation relationship between the relative heat generation of the low-temperature part of the RS of the amorphous alloy and the condition of the material after low-temperature annealing with the parameters: $T_a = 553$ K, $t_a = 100$ h, confirming by experiments the ductile–brittle transition criterion (5.19).

this book to qualify failure of at least one specimen of the set ($1 < N < 15$), where N is the total number of the samples in the set.

Equation (5.20) was used to determine the time to the start of the ductile–brittle transition in the amorphous alloys after low-temperature annealing (Fig. 5.6). The upper integration limit was determined by projection on the temperature axis of the point taken at half width of the left branch of the high-temperature part of the RS and equalled $T = 668$–673 K. Figure 5.6 shows clearly that if the ratio of heat generation $H(T_a, t)$ of LTRS $\int h(T_a = 433\,\mathrm{K}, T, t_0 = 100\,h)\,dT \int_T h(T, 0, 0)\,dT$ is a positive value, the material of the amorphous alloy is in the ductile state, if it is negative – in the embrittled state. The point of intersection of the ratio $\int_T h(T_a, T, t)\,dT \int_T h(T, 0, 0)\,dT$ with the t axis, where t is the annealing time, determines the time to the start of the ductile-brittle transition. The above-mentioned ratio of the amorphous alloy (e) intersects the axis ln (t) at a point which coincides almost completely with the time to the start of the ductile–brittle transition (98 h). It is evident that the amorphous alloy (e) should after annealing show the 'semi-embrittled' behaviour in the temperature–time range similar to that shown in Fig. 5.5. The relationship (5.20) was used to predict the stability of the mechanical properties of the amorphous alloy on the basis of the change of the spectral density of heat generation LTRS of the amorphous alloy after holding at room temperature ($T = 293$ K, $t_0 = 8000$ h). The results are presented in Table 5.2

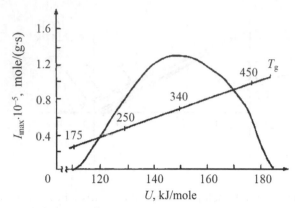

Fig. 5.7. Functional dependence of the maximum heat generation I_{max} on the activation energy U obtained for the amorphous alloy (e); T are temperatures corresponding to the activation energies U.

together with the parameters of thermal and time stability obtained independently using the T_{br}-criterion.

The spectrum of the activation energies was determined from the spectral density of heat generation of the RS for a typical amorphous alloy of the system (e). The temperature dependence $h(T_a, T, t)$ shows that its peak is in the range $T = 618–628$ K, and it is therefore accepted that $T = 623$ K. The temperature displacement of the heat generation peak is determined by the variation of the heating rate ($T = 5, 20, 50$ K/min). The equation (5.17) was used to calculate the activation energy corresponding to the heat generation peak of the RS and was equal to $U = 148$ kJ/mole · deg, and the value of the parameter $x^* = 29.54$ was also determined. The combined solution of (5.16) and (5.17) with the previously determined value of the parameter x^* gave the adjacent values of the activation energies and I_{max} (U) (Fig. 5.7). The figure shows clearly that the freshly quenched amorphous alloy is characterised by the activation energy in the range 120–180 kJ/mole ·deg.

Thus, summarising the above results, it can be concluded that there are two independent, physically substantiated criteria for the evaluation of the time to the start of the ductile–brittle transition of the amorphous alloys in the climatic temperature range. To obtain T_{br}-criterion of the ductile–brittle transition it is necessary to evaluate the temper brittleness temperature of the alloy in the quenched state and after annealing at the given thermal and time parameters. To apply the RS criterion, it is necessary to use the DSC method to obtain two heat generation spectra, one of which corresponds to the quenched state of the alloy and the other one to the condition after annealing at the given thermal and time parameters.

Both criteria predict similar values of τ_0 at room temperature: 40–45 years for $Fe_{46}Co_{32}Si_9B_{13}$, 8.5–9 years for the $Fe_{76}Ni_2Si_9B_{13}$ alloy and 25.5–27 years the $Fe_{70}Cr_{15}B_{15}$ alloy. The addition of phosphorus to the $Fe_{46}Co_{32}Si_9B$ and $Fe_{70}Cr_{15}B_{15}$ alloys greatly reduces the time to the start of the ductile–brittle transition in them: to 8–9 and 6–7 years for the $Fe_{70}Cr_{15}P_{12}B_2Al_1$ alloy [6, 7].

References

1. Zaichenko S.G., et al., Restoring the mechanical properties of amorphous alloys after embrittling annealing. Chernaya metallurgiya. Bulletin of Science and Technology Information. Moscow: Chermetinformatsiya - 1987 - No. 16. - P. 55-56.
2. Glezer A.M., et al., Structural causes of temper embrittlement of amorphous alloys of metal–metalloid type. FMM. - 1984. - V. 58. - No. 5. - P. 991-1000.
3. Kissinger H.E. Reaction kinetics in differential thermal analysis. Analyt. Chem.- 1957. - V. 29. - No. 11. - P. 1702-1706.
4. Ozawa T. Kinetic analysis of derivative curves in thermal analysis. J. Thermal Analysis. - 1970. - V. 2. - No. 3. - P. 301-324.
5. Glezer A.M., et al. Physical prediction criteria of ductile–brittle transition in amorphous alloys. FMM. - V. 80. - No. 2. - P. 142-152.
6. Glezer A.M., et al., Ductile-brittle transition and temperature and temporal stability of amorphous alloys. Izv. RAN. Ser. Fiz. - 2015. - V. 79. - No. 9. - P. 965-972.
7. Glezer A.M., et al., On the question of the physical criteria for time-temperature stability of the mechanical behavior of amorphous alloys. Deformatsiya i razrushenie materialov. 2015. - No. 3. - P. 2-7.
8. Kimura H., Masumoto T., Strength, ductility and toughness - consideration within the mechanics of deformation and fracture. Amorphous metallic alloys. - Moscow: Metallurgiya, 1987. - P. 183-221.
9. Zaichenko S.G., Glezer A.M. Izv. RAN. Ser. Fiz. - 2004. - V. 68. - No. 10. - P. 1488-1494.
10. Filippovich V.N., Kalinin A.M., The Glassy State. - Leningrad: Nauka, 1969.
11. Lavrent'ev M.A., Shabat B.V., Methods of the theory of functions of complex variables. - Moscow: Nauka, 1964.

6

Methods for increasing thermal and time stability

Taking into account the restrictions on the heat treatment of the amorphous alloys which are introduced by the ductile–brittle transition (temper brittleness), it is very tempting to apply these effects to the amorphous matrix which would reduce the rate of the process of coalescence of submicropores and would result consequently in an increase of the temperature threshold of embrittlement which in fact means the increase of the thermal and time stability of the amorphous state of the alloys.

So far, these effects included thermomechanical [1, 2], ultrasound [2, 3] and radiation [4] treatments. Thermomechanical treatment is based on the application of a constant load creating uniaxial elastic stresses at temperatures considerably lower than the glass transition point of the amorphous alloys but slightly higher than room temperature. In [1] ribbons of the Fe–B, Fe–Ni–B and Fe–Cr–B amorphous alloys and also cobalt-based alloys were annealed under stress. The alloys were heated in a vacuum furnace at a rate of approximately 100°/h. The one side of the ribbon was rigidly secured and on the other side there was a weight suspended through a system of blocks generating uniaxial tension in the range $(0.05–0.30)\sigma_T$, where σ_T is the yield stress of the amorphous alloy at room temperature. Annealing for 0.5–3.5 h was carried out in the temperature range $(0.1–0.3)T_{cr}$, where T_{cr} is the crystallisation temperature of the amorphous alloy. The load was removed after cooling the specimen to room temperature at the rate identical with the heating rate.

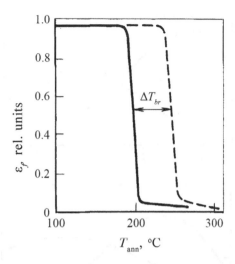

Fig. 6.1. Dependence of ductility parameter ε_f on the preliminary annealing temperature T_{ann} for the Fe–B alloy without thermomechanical treatment (solid line) and with thermomechanical treatment (dashed line).

Figure 6.1 shows the $\varepsilon_f(T_{ann})$ dependence for the Fe– B alloy annealed under stress in the conditions 0.15 T_{cr} for 1 h at $\sigma = 0.1\sigma_T$ and not subjected to preliminary thermomechanical treatment. It may be seen that the large decrease of the ductility in the alloy after preliminary treatment takes place at a higher annealing temperature. In other words, annealing under stress displaces the value of T_{cr} to higher temperatures. Figure 6.2 shows the increase of the temper brittleness temperature ΔT_{br} with the variation of applied stress, temperature and duration of preliminary treatment for the Fe– Cr–B alloy and the cobalt-based alloy, respectively. It may be seen that the increase of T_{br} takes place mostly at the optimum values of the varied parameters because the dependences $\Delta T_{br}(\sigma)$, $\Delta T_{br}(T)$ and $\Delta T_{br}(\tau)$ in all cases are described by the curves with a maximum.

In [2] it was established that the combination of temperature and bending plastic deformation has a very strong effect. Figure 6.3 shows the variation of T_{br} with the duration of isothermal annealing for the $Fe_{55}Co_{26}B_{14}Si_5$ alloy at 180°C and for the $Fe_{70}Cr_{15}B_8P_7$ alloy at 120°C. Some of the ribbon specimens were annealed in the condition after coiling on to a rigid core with a diameter of 10 or 1.5 mm (coiled state). Some of them were tested in the condition without coiling. It may be seen that in the preliminary annealing of the specimens without coiling the value of T_{br} decreases at a constant rate in proportion to ln τ, where τ is annealing time.

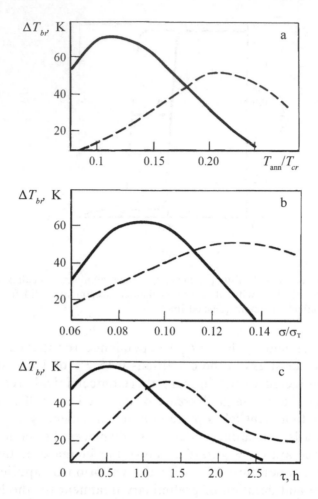

Fig. 6.2. Dependence of the increase of the temper brittleness temperature ΔT_{br} on the preliminary annealing temperature T_{ann} at $\sigma = 0.1\ \sigma_T$ and $\tau = 1h$ (a), on the applied stress σ at $T_{ann} = 0.15\ T_{br}$ and $\tau = 1$ h (b); on annealing time τ at $T_{ann} = 0.15\ T_{cr}$ and $\sigma = 0.1\sigma_T$ (c)): solid lines – Fe–Cr–B alloy; dotted lines – the cobalt-based alloy.

In cases in which heat treatment was carried out for the coiled specimens, the variation of T_{br} was non-monotonic: initially, there was a small increase and then a gradual decrease. Since the annealing of the coiled specimens was accompanied by plastic deformation and, consequently, the ribbon had a residual curvature, it should be assumed that the increase of T_{br} is caused by the simultaneous effect of temperature and plastic deformation.

As in the case of other destabilizing effects, there is the optimum degree of deformation at the given annealing temperature resulting

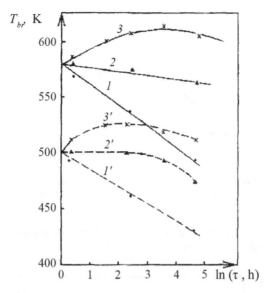

Fig. 6.3. Dependence of T_{br} on the annealing time for specimens without coiling (curves 1 and 1′) and also for the specimens coiled on a core with a diameter of 10 mm (curves 2 and 2′) and a diameter of 1.5 mm (curves 3 and 3′); solid line – $Fe_{55}Cu_{26}B_{14}Si_5$ alloy after annealing at 180°C, dashed line – alloy $Fe_{70}Cr_{15}B_8P_7$ after annealing at 120°C.

in the maximum increase of T_{br}. Figure 6.4 shows the dependence of the positive effect of T_{br} on the degree of plastic deformation in bending which is expressed by the parameter t/R (t is the thickness of the amorphous ribbon, R is the radius of the residual curvature of the ribbon after annealing the coiled specimens) for all the alloys investigated in [2] the positive effect is most distinctive at $t/R = (12–13)\cdot10^{-2}$. Comparison of the values of ΔT_{br} with the absolute values of T_{br} for the investigated alloys shows that the positive effect from annealing in the coiled state is evident for the alloys characterised by lower absolute values of T_{br}.

It is quite simple to apply the effect of increasing the thermal–time stability of the plastic properties of the amorphous ribbon alloys in practice and requires only preliminary experiments to determine the optimum values of the coiling radius, temperature and preliminary annealing time after which the value of the parameter t/R assumes the specified optimum value.

Figure 6.5 shows the change of ΔT_{br} for the Fe–Cr–B alloy after applying preliminary alternating stresses of the ultrasound frequency (20 kHz) not exceeding the yield stress at room temperature in dependence on the value the parameter $b = \sigma_0/E$, where σ_0 is the

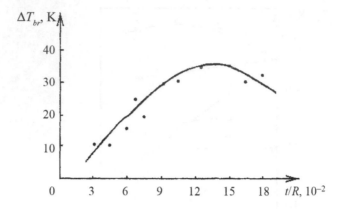

Fig. 6.4. Dependence of the increase of T_{br} on the degree of plastic deformation in annealing the coiled ribbon specimens.

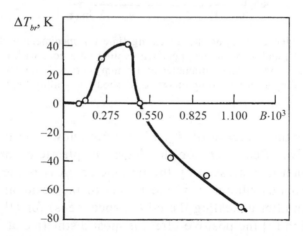

Fig. 6.5. Variation of the value ΔT_{br} in dependence on the value of the parameter $b = \sigma_0/E$ after applying preliminary alternating ultrasound oscillations with a frequency of 20 kHz (σ_0 is the amplitude of oscillations, E is the Young modulus).

amplitude of alternating stresses and E is the Young modulus [2]. The positive value of ΔT_{br} corresponds to the displacement of T_{br} to the high-temperature range, and the negative value – to lower temperatures. Here, there is a curve with a maximum: initially the value T_{br} is displaced to higher temperatures, reaching a change of approximately 50°C, and then with increasing amplitude of ultrasound oscillations the strength of the effect decreases. This is followed by the transition to the range of negative ΔT_{br} which indicates in fact the embrittlement of the amorphous alloy under the effect of ultrasound. The optimum value of the parameter b is 0.3–0.4 but in principle the ultrasound oscillations can, in dependence on the amplitude, increase or decrease T_{br}.

To explain the nature of the effect of ultrasound on the temper brittleness parameters, in [2] it was attempted to estimate the activation energy of the process taking place in the amorphous matrix and leading to the displacement of T_{br} to higher temperatures. This was carried out by estimating the increase of T_{br} with the variation of the amplitude of the ultrasound effect at temperatures of 50, 100 and 150°C. Assuming that the Arrhenius law is satisfied, the activation energy of the processes taking place in the Fe–Cr–B amorphous alloy was estimated and was equal to 0.5–0.6 eV. The effects are identical with those observed in thermomechanical ultrasound treatment and were also detected in irradiation of the amorphous alloys.

In [3] it was proposed to develop a method of ultrasound effect which would make it possible to influence the thermal–time stability of relatively large volumes of the amorphous ribbon. Figure 6.6 shows the diagram of equipment designed for ultrasound treatment of amorphous ribbons in the continuous mode. The amorphous ribbon (1) was displaced using a system of two electric motors (2) along the guiding plate (3) with a given speed of v_{dis} = (0.07–14.4)·10^{-4} m/s with a given level of tension of the ribbon which corresponded to the static tensile stresses $(0.06–0.3)\sigma_T$, where σ_T is the yield stress of the amorphous alloy. Ultrasound equipment (4) was installed in such a manner that the gap between the oscillating end of the waveguide-emitter (5) and the surface of the amorphous ribbon, positioned on the plate (3), is equal to 0. In a number of experiments balls (6) of ShKh15 steel were placed between the end

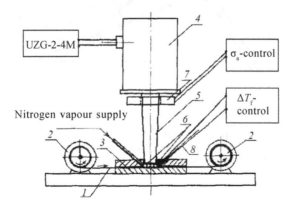

Fig. 6.6. Diagram of equipment for continuous ultrasound treatment of the ribbons of amorphous alloys (for notations see the text).

surface of the waveguide-emitter and the surface of the ribbon. The magnitude of the alternating stresses of the ultrasound frequency, excited in the amorphous ribbon, moving at a given velocity, was varied in the range σ_0 = 45.7–225.6 MPa and was controlled using an electrodynamic amplitude sensor (7). The temperature of the amorphous ribbons was maintained at the required level by regulating the supply of liquid nitrogen vapours and controlled using a thermocouple (8).

The results of the experiments carried out with the ultrasound effect on the temperature T_{br} for the $Fe_{75}Ni_2B_{30}S_{10}$ alloy are presented in Fig. 6.7. It may be seen that ultrasound treatment of the amorphous ribbon travelling at a given speed (v_{dis} = $1.3 \cdot 10^{-4}$ m/s) with a flat oscillating end of the waveguide-emitter increases the embrittlement temperature of the alloy in the entire investigated range of the amplitudes of alternating stresses σ_0 (curve 1). The dependence $\Delta T_{br}(\sigma_0)$ has the form of a curve with a maximum of ΔT_{br} at a specific value of σ_0. The form of the dependence $\Delta T_{br}(v_{dis})$ at a constant value of σ_0 = 45.7 MPa (curve 2) in the conditions of ultrasound treatment of the amorphous ribbons using the balls [2] is the same.

Thus, when using the proposed loading method the magnitude of the variation of the embrittlement temperature of the amorphous alloys under the effect of alternating stresses of ultrasound frequency depends on both the amplitude and the speed of displacement of the

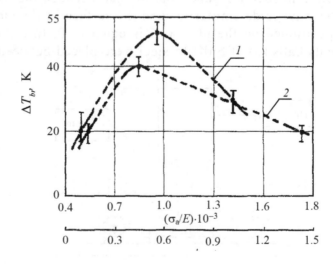

Fig. 6.7. Dependence of the increase of T_{br} on σ_0/E (curve 1) and on the speed of displacement of the ribbon in the treatment zone v_{cr} (curve 2); curve $1 - v_{cr}$ = 1.3 $\cdot 10^{-4}$ m/s; curve $2 - \sigma_0/E$ = 0.286.

amorphous ribbons, i.e., on the holding time in the zone of active ultrasound treatment. The application of the optimum parameters of treatment increased T_{br} by 40°C for $Fe_{75}Ni_2B_{13}Si_{10}$ alloy which in accordance with the T_{br}-criterion of the thermal–time stability, described in chapter 5, increases by 1.5 years the time resource of the ductility of the alloy at room temperature.

The authors of [5] carried out in irradiation with thermal neutrons of the $Fe_{40}Ni_{40}B_{20}$ amorphous alloy in the initial condition and after annealing above T_{br}. In this case, the alloy was transferred to the brittle state. Figure 6.8 shows the variation of the ductility in bend testing ε_f in dependence on the neutron flux acting on the specimen (annealing prior to irradiation above T_{br}). Starting at a flux of $2 \cdot 10^{17}$ cm^{-2}, plastification of the amorphous alloy takes place; the value ε_f increases to unity. This is in fact equivalent to the increase of T_{br} because annealing at this temperature no longer results in extensive embrittlement. As in the case of thermomechanical or ultrasound effects, there is the optimum value of the neutron flux increasing the ductility and equal to 10^{17}–10^{18} cm^{-2}. If the flux density is higher than 10^{18} cm^{-2}, embrittlement again takes place. Approximately the same neutron flux results in embrittlement of the initially ductile specimens not subjected to preliminary heat treatment.

In the investigations, the assumptions of the easier and peculiar occurrence of the structural relaxation in the conditions of tensile loading in the elastic region and also in the conditions of ultrasonic

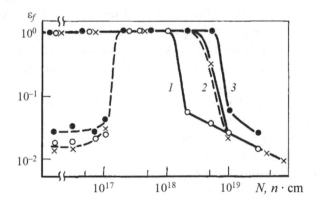

Fig. 6.8. Variation of the ductility parameter ε_f in dependence on the dose of preliminary neutron irradiation of the $Fe_{40}Ni_{40}B_{20}$ alloy. The thickness of the ribbon specimens, μm: curve 1 – 50; curve 2 – 40; the alloy was annealed prior to radiation above T_{br}.

oscillations or irradiation with high energy particles were taken into account. In this case, it is possible to propose several possible structural reasons for the effect of the previously investigated treatments on the value of T_{br}:

1. The decrease of the eccentricity of the regions of the free volume having the form of ellipsoids and, consequently, a decrease of the volume of the regions generating stress concentrations which in principle should increase the effective critical size of the micropores generating the cracks.

2. Collapse or fragmentation of the parts of the regions of the free volume with the form closest to that of the discs which are favourably oriented in relation to the applied exciting effects.

3. Ordering of the most mobile defects in the fields of elastic stresses. This may lead to the formation of a sort of superlattice with the nodes occupied by the free volume regions and consequently reduce the stimulus for their subsequent coalescence.

The possibility of the diffusion processes taking place in the presence of stresses leading to the disappearance of micropores results directly from the theoretical considerations [6]: large microirregularities should as a result of self-diffusion 'floating' disappear at a high rate than smaller microirregularities. In addition to this, in [7] it is concluded that the free volume is not capable of the instantaneous redistribution and requires a certain time to obtain the new equilibrium distribution which would result in the equalisation of the microstresses in the volume of the amorphous matrix. As regards the formation of the lattice of the micropores, the lattice was observed in the alloys subjected to irradiation [8].

The activation energy of the process leading to an increase of T_{br}, obtained in [3], is similar to that which, according to the estimates by a number of authors, corresponds to the activation energy of the process of migration of the free volume of a relatively small size. In early stages of the effects the migration of the regions of the free volume creates a structural state resulting in the slow development of the process of formation of microirregularities of the critical size. At a later stage of the external effects the migration of the free volume creates suitable conditions for high rates of formation of pores which, in the final analysis, either do not change the value of T_{br} or reduces this value. This applies in particular to the ultrasound or radiation effects in which the formation of micropores in the stage of rapid decrease of T_{br} can be observed by optical and electron microscopy.

Thus, temper brittleness is a structure-sensitive characteristic of the amorphous alloy and can be stimulated or, on the other hand, inhibited by influencing in the corresponding manner the regions of the potential nucleation of quasi-brittle cracks in the amorphous matrix. The above described effect of the annealing under stress and also the effect of ultrasound oscillations or irradiation with high-energy particles on T_{br} is another confirmation of the relaxation nature of the temper brittleness of the amorphous alloys. In addition, this effect confirms the controlling role played by microirregularities in embrittlement. These microirregularities are a direct consequence of the existence of the free volume in the amorphous alloys.

The thermal and time stability may be positively affected by the deposition of thin crystalline layers on the surface of ribbon specimens of the amorphous alloys. In [9, 10] coatings of nickel, titanium and barium titanate were deposited by vacuum spraying in the Plazmennyi kotel equipment in the pulsed mode in which the specimens could not be heated to temperatures exceeding 353–373 K. The coating thickness was 100 nm. The experimental results show that the maximum effect on the increase of the thermal and time stability of the mechanical properties of the amorphous alloys is exerted by the coating of chemical purity nickel resulting in a large increase of the temperature of the ductile–brittle transition in low-temperature annealing.

References

1. Glezer A.M., et al., Annealing and temper brittleness of amorphous alloys, FMM. - 1988. - V. 65. - No. 5. - P. 1035-1037.
2. Aldokhin D.V., et al., The study of the transition conditions of amorphous alloys from the plastic to the brittle state. Vestn. Tambov. Gov't. Univer. Ser. Estest. Tekh. Nauk. 2003 - V. 8 - No. 4. - P. 519-521.
3. Glezer A.M., Smirnov O.M. Effect of ultrasound on temper embrittlement of amorphous alloys, FMM. - 1992. - V. 68. - No. 34. - P. 1411-1612.
4. Gerling R., Schimznsky F.P., Wagner R. Influence of the thickness of amorphous $Fe_{40}Ni_{40}B_{20}$ ribbons on their mechanical properties under neutron irradiation and thermal annealing, Proc. Fifth Int. Conf. RQM. Elsevier Sci. Publ. - 1985. - V. 2. - P. 1377-1380.
5. Skakov Yu.A, Finkel' M.V. Izv. VUZ. Chernaya Metallurgi. - 1986. - No. 9. - P. 84-88.
6. Lirikov L.N. The healing of defects in metals. - Kiev: Naukova Dumka, 1980.
7. Taub A.I. Measurement of the microscopic volume strain element in amorphous alloys, Scr. Met. - 1983. - V. 17. - No. 7. - P. 873-878.
8. Gurovich B.A., et al., Fiz. Met. Metalloved. - 1981. - V. 61. - No. 4. - P. 922-930.

9.   Aldokhin D.V., et al., Influence of destabilizing impact on the mechanical properties of amorphous alloys. Materialovedenie. - 2004. - No. 7. - P. 28-36.
10.   Glezer A.M., et al. Physical prediction criteria of the ductile–brittle transition in amorphous alloys, FMM. -V. 80. - No. 2. - P. 142-152.

Conclusion

The originality and uniqueness of the physical properties of amorphous alloys is determined by the specific features of their structural state. Since the middle of the 70s of the previous century, special attention has been paid to the structure of amorphous alloys and a number of important results have been obtained. However, even at the present time, the structural states of the amorphous alloys are far from being completely understood.

Already the first X-ray studies determined the absence of the translational symmetry in the amorphous alloys and the fact that their structure is close to the structure of the liquid. Naturally, the term 'amorphous state', like the term 'crystalline state', assumes the existence of a wide range of different structures formed depending on the method of production, chemical composition and subsequent treatment.

Of course, the amorphous states of the metallic systems are usually highly non-equilibrium. Even small changes in the conditions of existence (temperature, load, composition, etc) may result in large changes of the structural and phase states and the properties. In other words, the amorphous state as the low-stability state of the condensed system may change even in the case of small changes of the external conditions. However, the controlling parameters (the parameters which determine the structural–phase state and, consequently, the properties of these systems) differ for different systems. It is not always clear which parameters should be changed in order to obtain the required physical and mechanical properties of the material.

The condensed systems, undergoing phase and structural transformations, are interesting because of the fact that, in particular, their structure in the region of the transitions has the special features typical of the transitional processes. It is well-known that the structural–phase transformations are one of the transformations that are most difficult to describe.

Therefore, the fundamental and, in particular, physical aspects of the formation, behaviour and special features of the structure and properties of the amorphous states of the condensed system with low stability with respect to the external conditions are the main subject of this book.

Structure

Various models of the structure of the amorphous alloys are characterised by the construction algorithms (models based on the random dense packing of the atoms, cluster and polycluster models, models based on distorted spaces, disclination and dislocation–disclination models), the selection of the interatomic interaction potential and the methods of minimising energy or stress relaxation. The models of amorphous substances are constructed using methods of modelling atomic systems, such as the molecular dynamics methods, the statistical relaxation methods, the Monte Carlo method, the methods used to construct the models on the basis of the available diffraction data for the structure, etc. The structural models of the amorphous alloys can be divided into two large groups: the first group – the models based on the quasi-liquid description of the structure using a continuous network of randomly distributed atoms; the second group – the models in which examination is based on the quasi-crystalline description of the structure by means of clusters or crystals containing a high density of defects of different type. Subsequent attempts to construct the models of the structure of the binary amorphous alloys can be divided into two main directions:

– computer construction using the model of the random high density packing of rigid spheres (RHDPRS) of the structure which would be subsequently subjected to relaxation using the appropriate potentials of the paired atomic interactions. The final structure should in this case describe accurately the main special features of the general radial distribution function (RDF). Although the algorithm of construction of this model for metal–metalloid type alloys requires the minimisation of the number of the nearest neighbours of the metalloid–metalloid type, the resultant number of these bonds differs greatly from zero;

– construction of the 'stereochemical' models in which the clusters, consisting of the atom-metalloid and the metal atoms surrounding it, form together the coordination cell (for example, in the form of a trigonal prism). In this case, the binary alloys of different compositions are studied as a simple mixture of close-packed regions of the pure metal and the regions with the structure with dense packing of trigonal prisms in the vicinity of the atoms-metalloids. Analysis of the applicability of the stereochemical model shows that it enables an accurate description of the structure of the amorphous alloys of the transitional metal (TM)–metalloid (M) type at high metalloid concentrations. In all stereochemical models, the non-periodic packing of the atoms relates to the absence of the periodic potential of the lattice so that the existence in the structure of the defects of the type of vacancies and dislocations becomes problematic, but the concept of the inter-cluster boundaries is introduced.

Structural defects

To understand the role of the defect in a specific process it is necessary to consider in particular the state of the structure without defects. Comparison of the state with the defects and without them can be carried out in terms of the topological properties or in terms of the stress fields. The majority of defects typical of crystals lose their specific features in the amorphous state. Nevertheless, the results of a large number of experiments carried out to examine the structure-sensitive properties of the amorphous alloys show that the structural defects can also exist in amorphous alloys. The deviation in the structure of the amorphous solids from the low-energy equilibrium state may be described by the increase of the density of these defects. Taking this into account, it is not possible to determine as the defect some small ordered region in the amorphous matrix, although the small disordered regions in the crystal can be regarded as clusters of elementary defects.

Several attempts have been made to provide a generalised definition of the structural defects in amorphous solids. Initially, a model of the ideal amorphous structure was constructed, and subsequently, defects were introduced into this structure on the basis of the purely geometrical considerations by analogy with what is done with the crystals. This was followed by measurement of the resultant displacements which, usually, were very large. A similar definition of the defects, using the local deformation considerations, is regarded as acceptable only for covalent amorphous solids.

A number of investigators have sub-divided defects of amorphous alloys into internal and external. The former are typical of the materials even after extensive relaxation, and the latter – annihilate in the process of relaxation changes in the structure. Since in experiments it is very difficult to obtain any detailed information for the different types of defects and their distribution, it is necessary to use for these purposes the methods of computer simulation which provide data on the defects and on their evolution in the process of different external effects. In addition to this, these investigations will undoubtedly help in understanding the amorphous state.

The defects in the amorphous alloys can be subdivided into point, microscopic elongated and macroscopic. The main point defects, existing in the amorphous matrix, are: broken bonds; irregular bonds; pairs with the changed valency; the atoms with a small stress field (quasi-vacancies); the atoms with a large stress field (quasi-implanted atoms). The elongated defects include: quasi-vacancy dislocations; quasi-implanted dislocations; the boundaries between two amorphous phases; inter-cluster boundaries. The macroscopic defects include pores, the cracks and other macrodiscontinuities. An important source for the formation of structural defects of the amorphous alloys is the free volume determined in fact by the high expansion coefficient of the liquid.

The free volume can be regarded as having the form of cavities of a specific size or as a formation continually distributed in the matrix. As regards the network models of the structure of the amorphous solids (examining the continuous, mutually penetrating networks of the atoms) the positions in which the two networks are bonded by the formation of combined atoms are referred to as configuration traps. It is assumed that they form as a result of relaxation which is accompanied by the formation of a vacancy in one of the networks. The distributed volume of the vacancies is stationary because it is always associated with the already relaxed structure, existing around the stationary configuration. However, thermal activation may lead to the dissociation of a trap and disappearance of the vacancy which is displaced to another area, forming a new configuration trap.

Plastic deformation

The most surprising property of the amorphous alloys is their capacity for plastic flow. It has been assumed that plastic deformation is identical with the nucleation, multiplication and annihilation of dislocations, moving in the solid. However, there is no translational symmetry in the amorphous solid and, consequently, no dislocations in the classic understanding of this defect. Consequently, the amorphous solid should be absolutely brittle. In particular, this situation is typical of inorganic glasses, although there are also signs of a slight plastic flow. However, there is no plastic deformation in the amorphous alloys. In this case, it is possible to achieve the anomalously high strength which should be obtained in non-crystalline solids under the condition of preventing in them brittle fracture at stresses considerably lower than the yield stress. It should be noted that the capacity of the amorphous alloys for the plastic flow (by which they differ from other amorphous solids) is obviously associated with the collectivised metallic nature of the interatomic bond resulting in a considerably easier process of collective atomic displacements.

As in the crystals, depending on the extent of reversibility with time, the deformation in the amorphous alloys can be divided into elastic, inelastic and plastic. Elastic deformation is completely and instantaneously reversible after removing the load, inelastic deformation is completely reversible with time and, finally, plastic deformation is irreversible in time after removing the external load. In turn, the plastic deformation in amorphous alloys can take place by different mechanisms: homogeneous or inhomogeneous. In homogeneous plastic deformation, every element of the solid undergoes plastic shape changes because the uniformly loaded specimen undergoes homogeneous deformation. In inhomogeneous plastic deformation, the plastic flow is localised in thin discrete shear bands and the remaining volume of the solid is not deformed.

Homogeneous deformation in the amorphous alloys greatly differs from the viscous flow in the crystals. The main difference is that in the amorphous alloys the stage of steady creep is not reached and the strain rate continuously decreases with time. At the same time, the effective viscosity continuously increases. Another large difference is the fact that the similar increase of the viscosity takes place with time in accordance with a linear law, in both the freshly quenched and relaxed states.

This sort of behaviour of the amorphous alloys in homogeneous viscous flow indicates that the flow reflects in fact the processes of structural relaxation and it is accompanied by the continuous change of the structure. On the one hand, this assumption indicates that the atomic mechanism of the viscous flow and the structural relaxation is identical. On the other hand, it is necessary to carry out experiments with the viscous flow of amorphous alloys in the 'pure form' when the changes in the structure during deformation do not take place. A similar viscous flow, referred to as isoconfiguration, can take place if preliminary annealing is carried out at a temperature slightly higher than the test temperature. In this case, the irreversible change in the structure can be completely ignored.

The investigation of the change of the flow mechanisms is a relatively complicated experimental task. The direct structural and metallographic studies of investigation in this case provide only a small amount of information and in fact no answer. A suitable answer can be obtained by using indirect methods, which include the measurement of the non-isothermal recovery of the shape of the pre-deformed specimens and acoustic emission measurements in the loading process. The latter method is more convincing and provides more information.

Fracture

As in the crystals, the fracture of the amorphous alloys can be brittle or ductile. In the former case, fracture takes place by cleavage without any features of macroscopic flow at the stress lower than the yield stress. In uniaxial tensile loading brittle fracture takes place by the rupture of opposite faces, situated normal in relation to the tensile loading axis. Ductile failure in amorphous alloys takes place either after or simultaneously with the plastic flow process, and the material shows in this case the features of macroscopic ductility. In this case:

1. Fracture takes place on the planes of the maximum cleavage stresses;

2. Fracture is always associated with one (sometimes be to) transitions from one plane of the maximum cleavage stresses to another;

3. The fracture surface contains two characteristic zones: almost smooth cleavage areas and regions forming a system of interwoven 'veins'. The latter have the thickness of the order of 0.1 μm and are usually characterised by the ratio of the height to thickness from 2 to 4, with the exclusion of the points of the triple junctions.

The nature of all the previously described phenomena, associated with the special features of the fracture of the amorphous alloys, can be understood using the fracture model. The model is based on the controlling role of the free volume quenched from the melt and its interaction with the elements decreasing the surface energy. In fact, in any amorphous alloy in the initial condition there is a specific number of the free volume regions and its specific size characteristic distribution in the amorphous matrix depends on the composition of the alloy and the production conditions of the alloy.

Crystallisation

Of considerable interest is the variation of the mechanical properties in transition from the amorphous to the equilibrium crystalline state. The point is that there are situations associated with partial transition from the highly non-equilibrium amorphous state to the thermodynamically more stable equilibrium state under the effect of temperature, time and mechanical treatment, irradiation and other external factors which on its own is very important. In this situation, the amorphous–crystalline state can demonstrate a qualitatively new level of the mechanical properties. The amorphous state is characterised by low elastic moduli. Partial crystallisation evidently increases the value of this characteristic and creates suitable conditions for more effective application of the amorphous alloys in cases in which the material is subject to the requirements of not only high-strength and ductility but also relatively high values of the Young and shear modulus. The presence of the crystalline phase in the amorphous matrix may also cause undesirable consequences: the loss of ductility, structural heterogeneity, the appearance of local stresses, defects, etc. Evidently, much of this is determined by the conditions of the formation of the crystalline phase because this determines the morphology, phase composition and amount of the structural components in the amorphous – crystalline state.

The process of transition from the amorphous state of the crystalline state can be regarded as the order–disorder phase transition. This can take place in principle in heating of the amorphous state during cooling from the melt with the rate close to critical. In the first case, the crystallisation process takes place in the conditions of constant heat supply (either at constant or continuously increasing temperature) and with the additional effect of the heat generated during crystallisation. Consequently, the heat treatment of the system is accompanied in most cases by the formation in a specific stage of the structure consisting of two distinctive structural components: amorphous and crystalline. In this case, the nature of the structure depends on the heating and subsequent cooling rate, the heating temperature and the annealing medium. A completely different morphological type of the structure can form in different stages of crystallisation in the conditions of rapid cooling of the melt with efficient heat removal. Similar amorphous–crystalline formation have been investigated only in a small number of cases, but the mechanical properties, obtained in this case, can be regarded as completely unique.

Finally, there is another method of the formation of the amorphous–crystalline structure when dispersed crystalline particles of the refractory compound (usually, the carbide of a refractory metal) are added to the 'puddle' of molten metal formed on the quenching cooling disc. Consequently, the amorphised melt and, subsequently, the amorphous matrix, contains the particles of the crystalline phase, uniformly distributed in the volume.

Thus, the co-existence of the amorphous and crystalline phase should result in an increase of ductility. As soon as the amorphous matrix is crystalline, the ductility becomes closer to 0. As a result of the analysis of the reasons for the plasticising effect, it has been concluded that the interphase boundaries of the amorphous and crystalline phases can become sources of the free volume under specific conditions essential for the efficient currents of the processes of plastic deformation. It is interesting to note that identical effects can also be obtained in the mechanical mixture of a corundum powder and the appropriate lubricant.

Structural relaxation
The relaxation process of the stress state of the amorphous alloys has many common features with the extensively examined process of relaxation in the amorphous polymers and oxides but it influences the amorphous alloys to a considerably greater degree and in a larger number of physical and structural characteristics. In this case, it is not rational to discuss the properties of the amorphous alloys without taking into account the relaxation, because the measured properties can greatly depend on the structural relaxation of the specific state.

Homogeneous relaxation, referred to previously as structural relaxation (SR) takes place homogeneously in the entire volume of the specimen without influencing its amorphous nature. The process of SR is accompanied by the change of the short-range order leading to a relatively small decrease of the degree of non-equilibrium. In this case, the unstable atomic configurations, formed during amorphisation in the course of quenching, change to the stable configurations by means of small atomic displacements. This results in the densening of the amorphous matrix associated with partial annihilation and removal of the excess free volume. It is important that the displacement of the atoms during structural relaxation is smaller than the atomic spacing and takes place in local areas. The heat of transformation to the stable phase, which can be used as a measure of these non-equilibrium nature, changes only slightly in this case. The process of SR is accompanied by changes of many physical properties of the amorphous alloys: heat capacity, density, electrical resistance, internal friction, elastic constants, hardness, magnetic characteristics (changes of the Curie temperature, induced magnetic anisotropy), corrosion resistance, etc. A number of models have been proposed to describe the SR processes, and these models can be divided into two groups: 1. The models of the activation energy spectrum; 2) the model proposed by van den Beukel et al..

In the first model, it is assumed that the SR is caused by the local atomic rearrangement in the amorphous material taking place with different relaxation times (activation energies). The fundamentals of the models were described by Primak, and later applied to the relaxation processes in glasses. It is assumed that the activation energies of these processes are distributed in a continuous smooth spectrum. The rate of variation of the physical properties is proportional to the rate of variation of the density of 'kinetic processes'.

The second model uses the approach to describing the SR based on the classification of the short-range order, proposed by Etami. It is assumed that the first relaxation stage is accompanied by the compositional (chemical) short-range ordering. This contribution is efficiently described by the AES model and represents a reversible process taking place with the activation energy spectrum in the range from 150 to 250 kJ/mole. Chemical ordering is quite high speed and when it is completed the controlling factor is the topological short-range order. The topological relaxation is described by the free volume Spaepen model with the unit activation energy of approximately 250 kJ/mole and is an irreversible process. It should be stressed that the main assumption in the Spaepen model is that as a result of the heterogeneity of the structure of the glass it is accompanied by the formation of the areas with the excess free volume in relation to the 'ideal structure' – 'relaxation centres'. They may be characterised by thermally activated displacements of the atoms causing the redistribution of the free volume inside the material and, consequently, by its partial transfer to the free surface.

Because of the obvious shortcomings (the fact that it is not possible to determine directly the magnitude of the excess free volume, the presence of the unit activation energy of the relaxation processes) the Spaepen model of the free volume is used only in a small number of cases. The van der Beukel model is also being criticised because it is difficult to imagine the chemical and topological ordering of the processes independent of each other and taking place at different periods of time. At the present time, the models of the activation energy spectrum are used in most cases. In particular, the model of directional structural relaxation, i.e., the relaxation oriented by the external stress, is used widely.

It should be stressed that the amorphous structures (AS) are characterised by a very rare combination of the properties in the structural relaxation stage: unique mechanical characteristics (high strength, hardness, fracture energy, wear resistance), high corrosion resistance, useful magnetically soft properties), efficient machinability of the material. In addition, the AS are characterised by the elimination of the problem of producing the required foerms of the bulk semi-finished producted ands components based on them with isotropic properties in the cross section. All these advantages cause that they can be used as precision materials; magnetically soft materials; fibres or matrices of composite materials; in the manufacture of cutting instruments, sporting tools, mobile telephones, surgery instruments, prosthesis, etc.

As a result of the application of the new technologies of material processing (laser, plasma, chemical-thermal, magnetic pulse, treatment with high density currents, etc.) it is possible to stimulate the processes of structural relaxation and obtain the required combination of the surface and volume characteristics of the appropriate components made of the AS, a smooth transition of the properties in the cross-section (gradient materials) and synthesis of different structures. Special attention should be given to the method of high plastic strains. It has been shown that the torsion under high quasi-hydrostatic pressure of the amorphous alloys based on iron, nickel and titanium results in the self-blocking of the shear bands and delocalisation of the plastic flow and, consequently, in the nanocrystallisation and a large increase of the mechanical and magnetic properties.

The physical nature of the ΔT-effect

The amorphous alloys, produced by melt quenching, are in the metastable equilibrium state. Any destabilising effect (temperature, pressure, deformation, radiation, ultrasound) is capable of causing reversible or irreversible changes of the topological and compositional short-range order and, consequently, influence the physical properties of the disordered state. The thermal effects are reduced usually to the heating of the amorphous alloys above room temperature but lower than the solidification temperature of these alloys and, consequently, this results in the processes of structural relaxation, accompanied by the extensive rearrangement of the structure and changes of almost all physical and mechanical properties. At the same time, there may be changes in the magnetic properties of the amorphous alloys under the effect of cryogenic temperatures, for example, low-temperature thermal cycling in liquid nitrogen (T = 77 K). It is assumed that these changes may be caused by the martensite-like phase transition at a large decrease of temperature.

The results of the systematic investigations carried out by the authors of the book show that the phenomenon of the irreversible change of the structure and physical properties of the amorphous alloys after completing the low-temperature treatment are of a general nature and are similar to a certain degree to all amorphous alloys. On the basis of the experimental results it may be assumed that the nature of the observed low-temperature ΔT-effect is associated, the authors believe, with the transition of the amorphous alloy under the effect of low-temperature thermal cycling to a new structural state with different short-range order parameters. Systematic investigations showed that the main factors, which determine the extent of changes of the structure and physical properties, are the low-temperature treatment parameters (temperature in cooling, duration and the number of cycles) and the composition of the alloys. The physical model of the low-temperature ΔT-effect is based in particular on the examination of the thermal conditions of the cooling process of the amorphous alloys.

Thus, the thermoelastic stresses may cause failure of the atomic complexes, and the oscillations of the sheet are the driving force of the drift of the atoms. The combination of these processes results in changes of the short-range and medium-range order in the amorphous state, causes its homogenising and, in the final analysis, stimulates the transition to the state of the new metastable equilibrium of the amorphous metallic system.

Index

Spaepen free volume model 44, 91
stereochemical model 6, 157

N

nanocrystallisation 51, 58, 59, 62

P

Peierls forces 19
phase
  B2-phase 31
  Frank–Casper polytetrahedral phase 6
polyhedrons
  Bernal polyhedrons 4
  Voronoi polyhedrons 4

R

radial distribution function 3, 4, 5, 12
random close-packed soft spheres (RCPSS) 4
random dense packing of rigid spheres ((RDPRS) 4
relaxation spectrum 56, 113, 125, 130, 134, 135, 137
Rieman–Christoffel curvature 7

S

shear bands 14, 16, 17, 18, 19, 20, 21, 23, 24, 25, 40, 41, 51, 52, 79, 81,
    108, 109, 110, 117, 121
structural relaxation 11, 12, 15, 17, 31, 39, 40, 43, 44, 45, 47, 50, 51, 52,
    72, 76, 81, 84, 88, 89, 94, 96, 97, 113, 126, 127, 130, 135, 151
superplasticity 15, 39

T

T_{br}-criterion 134, 142, 151
temperature
  Curie temperature 43, 70, 82
  equicohesion temperature 27, 77
temper brittleness 74, 75, 76, 77, 78, 79, 81, 83, 84, 85, 86, 87, 88, 89,
    92, 93, 94, 95, 96, 97, 113, 114, 126, 131, 133, 142, 144, 145, 146,
    149, 153
temper embrittlement 74, 126, 143, 153
theory
  disclination theory 7
  the theory of the free volume 10